SERBS and CROATS

Other Books by Alex N. Dragnich

Tito's Promised Land: Yugoslavia

Serbia, Nikola Pašić, and Yugoslavia

*The Development of Parliamentary
Government in Serbia*

*The First Yugoslavia: Search for
a Viable Political System*

Major European Governments
(coauthor)

SERBS and CROATS

The Struggle in Yugoslavia

ALEX N. DRAGNICH

A HARVEST BOOK
HARCOURT BRACE & COMPANY
San Diego   New York   London

To
Paul and Alexander

Requests for permission to make copies of any part of the work should be mailed to: Permissions Department, Harcourt Brace & Company, 8th Floor, Orlando, Florida 32887.

Library of Congress Cataloging-in-Publication Data
Dragnich, Alex N.
Serbs and Croats: the struggle in Yugoslavia/Alex N. Dragnich.
p.   cm.
ISBN 0-15-181073-7
ISBN 0-15-680663-0 (pbk.)
1. Yugoslavia — Politics and government.
2. Yugoslavia — Ethnic relations.   I. Title.
DR1282.D73   1992
949.702 — dc20   92-19786

Designed by Lori J. McThomas
Printed in the United States of America
First Harvest edition 1993
A B C D E

Contents

Maps

Note on Spelling and Pronunciation of Serb-Croat Words and Names

a — a as in father

c — ts as in mats

ć — ch as in rich

č — ch as in chalk

dj — g as in George

e — e as in pet

i — i as in machine

j — y as in yet

lj — li as in million

nj — ni as in dominion

o — o as in over

š — sh as in shawl

u — u as in rule

ž — j as in French *jour*

Preface

The Serbs and Croats are at the heart of the tragic drama that has been unfolding in Yugoslavia in recent years, but the roles of other ethnic groups cannot be ignored. To understand the complex subject that is Yugoslavia, it is necessary to consider the relations among South Slav ethnic groups in the context of the events that led them to create their common state, and their experiences as components of that state.

I was led to write this book primarily by the realization that the journalism produced in the wake of the collapse of Communist regimes in Eastern Europe, including Yugoslavia's, has been not only inadequate, but also often incorrect. Errors of fact and highly misleading interpretations were owing, I believe, mainly to a lack of knowledge of European history, especially that of the Balkans. Other writers, equally ignorant of this history, repeated misinformations and questionable interpretations, and thus added to the confusion.

My aim here is to help clarify and correct the background for the average interested reader who is often impatient with lengthy books, countless footnotes, and endless quotations. Should this short history create doubt and skepticism, I invite readers to consult the books listed in the bibliography, where I have also listed my own books that treat many of my subjects in greater detail, and in which I have documented my findings and explained my judgments.

In discussion of Yugoslav affairs over many years, I have learned a great deal from colleagues and friends in the United States and Canada as well as in Yugoslavia, but it would be impossible to give credit to them individually. I am grateful to them all. When I realize how much time I have devoted to the study of Yugoslav history and politics over the years, I realize how much I owe to the patience of my family, for which I am also grateful. In connection with this book, I owe special thanks to William Jovanovich for his interest, his many valuable suggestions, and his spirited encouragement. I am especially indebted to Dr. Srdjan Trifković, who in the shortest possible time read the manuscript and offered useful criticisms and suggestions. In the final analysis, of course, I am the only one responsible for what is in this work.

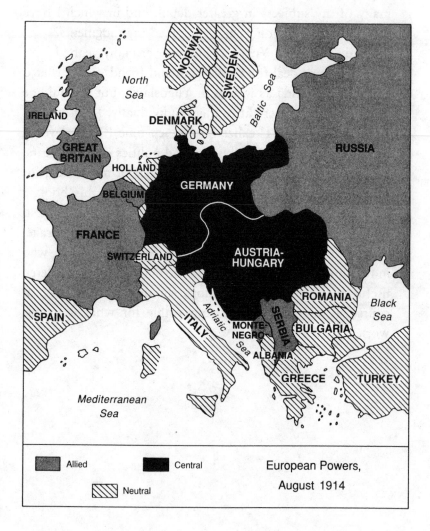

European Powers,
August 1914

Allied

Central

Neutral

Introduction

President Woodrow Wilson's espousal of self-determination in his Fourteen Points in 1918 kindled the flames of latent nationalism among many peoples in central and southeastern Europe. To the great powers of Europe this represented a disconcerting break with the past, because they had become accustomed to determining the fate of certain peoples, or at least to limiting their choices of action. Who were these "certain peoples"? They were either onetime nations that had been subjugated by empires, or they were distinct ethnic groups in more or less cohesive geographical areas which had never enjoyed independence. More precisley, the areas of southeastern Europe had been the sphere in which the great powers struggled for dominance, and in their struggles they sometimes "traded" one people for another in a callous game.

The competing interests of five empires—Ottoman, Russian, Austro-Hungarian, British, and French—served to prevent or hold back the peoples of the Balkans in their attempt to realize their national ambitions. These empires ignored, for the most part, the interests or aspirations of the local inhabitants and allowed them to flounder, educating some of them, favoring this region or that, to serve their own interests.

Ironically, the great powers accused local peoples of an inability to unite and a lack of understanding of what constitutes and maintains a state. Thus they justified their continued imperial presence and dominance. Moreover, they sought to mold world public opinion about events in the Balkans. Yet the local peoples continued to dream of freedom and independence.

The British Empire controlled no areas in the Balkans; its concern in that part of Europe was to maintain the security of its trade route through the Mediterranean to the Far East. Hence, it tended toward accommodation with the Ottoman Empire, which held domain over a large part of the Balkans. British opposition to Russia, as in the Crimean War, was to keep a balance of "great-power influence" in the approaches to the East.

The Russian Empire's major interest in the area was possession of a warm-water outlet to the Mediterranean. In opposition to the Ottoman Turks, it cast its motives for involvement in Balkan affairs in virtuous terms: concern for the protection of the Ottoman Empire's Christian subjects.

German interest in a Berlin-to-Baghdad railway worried the Russians. All the empires in Europe were involved in "eastering." Austria-Hungary not only incorporated Slovenes and Croats and many Serbs, but also determined to keep little Serbia weak. The French interest was similar to the British, but the French, in general, were motivated more to prevent the British or the Russians from gaining predominance in the area.

Empire building and subjugation of foreign peoples was

the prevailing situation at the beginning of the nineteenth century. But far-reaching changes were in the offing.

Ottoman power was on the wane, although it was to linger into the twentieth century. The Ottoman Turks had moved into southern Europe during the late Middle Ages, having subdued the Serbian state, once the largest empire in the Balkans, in two critical battles, Maritsa in 1371 and Kosovo in 1389. The fall of Constantinople to the Turks in 1453 marked the decisive establishment of their foothold in Europe. They pushed north and west and twice laid siege to Vienna, in 1529 and 1683, but they did not succeed against the Austrians and Hungarians in part because the Habsburg Empire had established the Military Frontier, populated largely by Serbs and others who had fled before the invading Turks. The Hungarians, particularly, gave these peoples land, freedom of worship, the right to elect their local leaders, and other privileges in exchange for their willingness to assist in fighting the Turks. There was a need for this, and it was relatively easy for the Hungarians to offer benefits to the Serbs, because the lands they offered had already been abandoned by Croats and others who had fled north.

The favorable conditions offered by the Habsburgs induced several large migrations of Serbs from their conquered homeland to the Military Frontier (*Krajina* in the Serbo-Croatian language). These migrations occurred during the sixteenth, seventeenth, and eighteenth centuries. The large Serbian majority in this area that resulted was to become a serious problem in 1941, when it was massacred by Croats, and in 1991, when Croatia proclaimed that it was seceding from Yugoslavia.

The Serbs could not remain subjugated forever. The

waning of Ottoman power was signaled by two Serbian uprisings, in 1804 and 1815. These led to the resurrection of the Serbian state. By mid-century, after international treaties were affirmed, Serbia gained de facto independence. Its formal international recognition occurred in 1878.

The Greeks had revolted against the Turks in 1821, and Montenegro, also Serb, was never wholly conquered by the Turks. It continued as a beacon of Serbdom in the Serbs' darkest days.

It is significant that the Russian tsar viewed himself as Montenegro's protector. The Serbs, including Montenegrins, were Orthodox Christians in religion and Slavic by blood. From time to time they were attached to the idea of Pan-Slavism, as were Czechs and others. That aspect of nineteenth-century geopolitics is not here a significant matter, however much it enters into the literature of the period, as in Dostoevsky.

With French expansion, Napoleon's liberalism opened minds in the Balkans to thoughts of freedom. The French occupied Dalmatia, including the republic of Dubrovnik, in 1806. And in 1809 they gained control of parts of the Military Frontier and all of the Croatian interior and Trieste, as well as most of the areas inhabited by Slovenes. These areas were consolidated into the Illyrian Provinces, which were made part of the French Empire. The French presence ended in 1813, but the national awakening it had fostered remained very much alive.

The collapse of the Napoleonic regime led to a period of instability throughout central and eastern Europe. One consequence was the revolutions of 1848. As they subsided,

Austrian and German tendencies toward unification began to grow.

The number of German states in 1800 was more than three hundred, but by the time of the Congress of Vienna (1815) they had formed a loose confederation of thirty-eight. The most influential were Prussia and Austria, which competed to unify the German peoples. In victorious wars against Denmark (1864) and against Austria (1866), Prussia was victorious. After its rapid defeat of France in 1870, unification of Germany was completed.

In 1867, Austria recognized Hungary as an equal partner to form a dual monarchy, which contained large numbers of South Slavs, Croats, Slovenes, and Serbs. Bent on gaining control of southeastern European lands that the decaying Ottoman Empire had been forced to abandon, it dominated or weakened independent Balkan states. For their part, the leaders in Berlin viewed Austria-Hungary's actions as paving the way for Germany's own expansion to the east, its ancient policy of *Drang nach Osten*. Among indications of the scope of Germany's ambitions in the nineteenth and early twentieth centuries was its proposal for a Berlin-to-Baghdad railway. Interest in the East seems fixed in the German mind, as when, in 1992, Germany pressured the European Community to recognize Slovenia and Croatia, though the United States held back.

Internationally, Austria-Hungary and Germany sought to secure their gains through several treaties. At the Congress of Berlin (1878), the Dual Monarchy gained the authority to occupy and administer Bosnia-Herzegovina. These two areas had earlier been under Turkish rule, and they were populated mainly by Serbs and other South Slavs. Russia agreed to the terms of the treaty mainly in order to obtain recognition of an enlarged Bulgaria. The Bulgarian gains were to be made at the expense of Serbia, in spite of Serbia's sacrifices on the side of Moscow in the wars of 1876 and 1877 against Turkey. Also at Berlin there was a barely secret understanding that at a later date Vienna would have the right to annex Bosnia-Herzegovina outright. The Russians understood that they would be consulted preceding such an event.

In further pursuit of their plans, the imperial leaders in Berlin and Vienna concluded a secret accord in 1879, to which Italy added its signature in 1882. This came to be known as the Triple Alliance and was avowedly a defense pact allying Germany, Austria-Hungary, and Italy. Earlier, Germany's Bismarck had sought to arrange a League of Three Emperors—German, Habsburg, and Romanov—but this had failed, mainly because of Austro-Russian rivalry in the Balkans. To counter the Triple Alliance, a British-French agreement was made shortly after the turn of the century, which gained a third partner, Russia, in 1907, to become the Triple Entente. By 1914, these three-nation alliances faced each other, ready for war.

War came in 1914 when a Serb patriot in Sarajevo killed the Habsburg heir to the throne. Austria-Hungary made extravagant demands on Serbia, which, by general agreement, no independent nation could accede to. Berlin reluctantly backed the Habsburg venture, and soon the Triple Alliance was at war with the Triple Entente. Soon Britain, France, Russia, Belgium, Serbia, Montenegro, and Japan were fighting Germany and Austria-Hungary. Italy refused to join the latter, arguing that since their actions were offensive and the

Triple Alliance was a defensive pact, there was no obligation to respond.

After nearly four years of brutal and indecisive trench warfare on the Western Front, the military situation at the beginning of 1918 looked grim. Italy, which had changed sides in 1915, had been invaded two years later, and the Bolshevik revolution in late 1917 was knocking Russia out of the war. The arrival of United States troops in 1918, together with failed new German offensives that year, signaled that the outcome of World War I was no longer in doubt.

Conveying confidence and idealism, President Wilson, in January 1918, announced his Fourteen Points as the basis of an honorable peace. Although he did not use the word "self-determination," his reference to the indigenous rights of peoples within Turkey and Austria-Hungary made his meaning unmistakable. Of the Balkan states, he mentioned only Serbia, Montenegro, and Romania by name (along with Poland, not then in existence independently).

Interestingly, Wilson did not mention Czechoslovakia, nor was there any reference to a combining of other Slavs with Serbs in Serbia and Montenegro into a Yugoslav state. The reason for this omission is simple: The victorious Allies had not yet agreed on the dissolution of the Austro-Hungarian Empire. Only when they did so, about mid-1918, was it realistic to think that peoples in the territories of the Dual Monarchy could attempt to form independent states.

Self-determination, however, was more meaningful as a concept to some peoples than to others. It would have been difficult to argue, for example, that it was as meaningful for the Slovaks or the Bosnians or the Slovenes as it was for the peoples of Serbia, Bulgaria, Romania, and Montenegro. The

Slovaks had never had an independent state, nor had the Slovenes; and the Croats were last a definable free entity in the eleventh century. It must be borne in mind, therefore, that some of the peoples of eastern Europe were parts of historically identifiable nations, whereas others were still seeking identities. Thus, some had experience in self-government and others had none.

Although the end of the Austro-Hungarian Empire opened all sorts of possibilities in Central and Eastern Europe, this book is concerned only with the experience of several South Slavs in seeking a common home in one state. Concretely, it is about those South Slavs—Serbs, Croats, and Slovenes— who in 1918 formed the Kingdom of the Serbs, Croats, and Slovenes. It was renamed Yugoslavia in 1929, *yug* meaning south.

Bulgarians are also South Slavs, but they continued with their separate identity, in part because there had been strife before World War I between Bulgaria and its neighbors. In 1913, Serbia, Montenegro, Romania, and Greece had defeated Bulgaria in the Second Balkan War. There are references to Bulgarians here, nonetheless, since they played a role in the history of the Yugoslav state. Likewise, there is some consideration of certain minorities that were constituent parts of that state.

Yugoslavia's existence, from 1918 to 1941, was shattered first by invasion and then by civil war. During World War II, Slovenes became part of Germany, and Croatia became a Fascist state. Following the war, Communists reestablished Yugoslavia, as a dictatorship under their party. In 1990, the country was again split by civil war. Yugoslavia now consists of Serbia and Montenegro. This book presents a treatment of the civil war that erupted during World War II, as well as an analysis of the Communist regime that followed. Finally, it endeavors to assess the Yugoslav experiment and its consequences, and raises the question of whether in 1992 the European powers, notably Germany, were seeking to impose a new settlement, just as they did at the Congress of Vienna in 1815 and the Congress of Berlin in 1878.

The Yugoslav Idea
and Its Antagonists

The beginning of the nineteenth century saw South Slavs under foreign rule, mainly Ottoman and Austro-Hungarian. The only exception was Montenegro, a small mountainous country populated by Orthodox Serbs.

The new century made a sharp break with the past. In 1804, the Serbs in Serbia revolted and embarked on the rocky road to regain their freedom. By mid-century they had gained de facto independence and began launching democratic political institutions. Thus, the Serbs of Serbia and Montenegro had their own states many decades before other South Slavs were liberated; the Croats and Slovenes, as well as a large number of Serbs, continued to live under foreign rule until after the second decade of the twentieth century.

The politically conscious South Slavs under foreign rule were aware that Serbia had regained independence and was striving for popular rule. Some of these other Slavs visited Serbia and discovered that they had a number of things in common with the Serbs, first of all language. Serbs and Croats speak the same language and write it the same except that Serbs use Cyrillic lettering and Croats Latin. The Slovenes speak a separate Slavic language, but it is a cousin to Serbo-

Croatian. Second, they were Christian, although the Serbs were mainly Orthodox, having received their Christianity from Byzantium; the Slovenes and Croats had received their faith from Catholic Rome. A significantly large number of South Slavs had converted to the Muslim faith, oftentimes on pain of death, during the long years of Turkish occupation. These Muslims were found mainly in the provinces of Bosnia and Herzegovina (usually called Bosnia-Herzegovina). In addition, there was in all Slavs who were subjugated, including Czechs and Slovaks and Poles, a sense of their common bondage and, in many aspects, a feeling of shared heritage. They had read the same Western writers who espoused liberty, and saw in this a more universal bond among them.

National stirrings among the South Slavs living under the rule of Hungary were hardly perceptible to the outside observer before the great year of revolutions, 1848. There had been little evidence of public organizational activity outside the free Slavic states of Serbia, Montenegro, and Bulgaria. Yet a few young Slovenes, Croats, and Serbs went to study in university towns, and there they picked up liberal ideas that were indeed radical for those times. These they carried back to their home areas, where the concepts proved infectious.

Foreign rulers, both Habsburg and Ottoman, had sowed seeds of distrust among the South Slavs, directed mainly against independent Serbia, which stood as a threat to their empires. On the other hand, the basic differences between western and eastern churches, although often recognized, did not seem to dampen the ardor of South Slav intellectuals bent on purveying a highly romanticized picture of a future common state.

It should be noted that there had been changes in religious affiliation among the South Slavs. Many Serbian Orthodox Christians, who at one time were strong in Dalmatia, notably in Dubrovnik, had become Roman Catholics. And many Serbs, as well as Croats, had converted to Islam during the long centuries of Ottoman occupation. Hence, confessional boundaries had tended to divide the South Slavs before the growth of national awareness. When that awareness came, the South Slav idea had to compete with Pan-Serbian and Pan-Croatian ideas.

Social differences were important too. Serbian society was peasant and patriarchal (but without bureaucratic superstructure) and relatively homogeneous. In Croatian society, class stratification was strongly pronounced, and allowed for little social mobility. The dominant voice was that of the town dwellers who had made accommodations with Vienna or Budapest. They promoted the study of Hungarian by making it a required subject in Croatian schools. Slovenian society, more industrialized than the other two, also had a differentiated class structure, one in which clericalism (Roman Catholic) and Germanization made great inroads. The German language was dominant beyond elementary school, and wealthy citizens were frequently foreigners.

In both the Croatian and Slovenian areas of the Dual Monarchy, the educational system sought to shape political attitudes by emphasizing how Croats and Slovenes differed from Serbs. Austria-Hungary had been rewarded by the Slovenes' and Croats' exemplary service in both civil duties and the military. The Croats especially had a remarkable record in Habsburg uniforms, particularly in the Imperial Navy, and were known throughout the empire for their fierce loyalty to the crown. The Serbs in the Military Frontier, the *Krajina,* also had good records of loyalty. They served as a buffer between the Turks and Habsburg lands.

These factors could not but have some impact on subsequent political outlooks, and they made for various complexities and difficulties for those who espoused the idea of a Yugoslav state. When specific proposals were made, these differences began to appear. Since Serbs lived in five of the South Slav territories (Serbia, Croatia, Bosnia, Herzegovina, and Montenegro) and the Croats in three (Croatia, Bosnia-Herzegovina, and Serbia), the idea of one state, as a factor, could not unite them, any more than religion, historical experience, or social values could. The one common denominator was language.

Language could not, however, reverse attitudes. For example, political leaders in Serbia looked upon Belgrade as the natural capital of all South Slavs, while those in Croatia and Slovenia viewed Zagreb as the natural center. The Slavs in Macedonia had been so embroiled in conflicts among Serbs, Bulgarians, and Greeks while suffering Ottoman rule that they lacked a nationalistic center.

Moreover, differing historical-political experiences affected the respective outlooks. Serbs distrusted authority. When they had fought for and won their independence, they established a constitutional parliamentary system. The peasants in Serbia had followed their local leaders in the wars against the Ottomans and continued the cooperation in popularly elected parliaments. The Montenegrins were, and are, a tribal people. Their theocratic state, not as autocratic as it might seem, was ruled by a tribe, the Njegoši, who had been chosen as leaders.

The Croatian and Slovenian peasants had not had such experiences. In the years when the Serbians were forcing their monarch—two royal families competed during the nineteenth century—to leave the throne, their counterparts in Croatia and Slovenia seemed powerless against local officials, let alone those at the top.

Although Croats and Slovenes did not have their own parliaments, they did have political groups or parties, in which, in certain areas, the Serbs among them also participated. These groups differed from those in Serbia, where ideas and programs influenced party alignments. In the Croatian areas of the Dual Monarchy political alignments tended to follow nationality lines. There was a Croat-Serb coalition in parts of Austria-Hungary, but it is significant that it was not thought of as a Yugoslav or South Slav coalition. Conversely, when the Serbs considered a free state of related and neighboring Slavs, they thought in terms of melting into a new nationality—Yugoslav.

Among the South Slavs the most vocal and uncompromisingly bitter antagonists of the South Slav idea were some extremist Croatian nationalists. The best known of these were Ante Starčević and Josip Frank, both active in the latter half of the nineteenth century. Starčević did not recognize as distinct other South Slavs; for him, they were all Croats, even if of a different religion. Even the Bosnian Muslims were Croat "blood brothers." The borders of his Croatia were essentially those of what became Yugoslavia. His idea of Great Croatia encompassed all the Serbs, though he refused to recognize the appellation "Serb," insisting that the name came from the word "Slavo-Serb," signifying a slave origin. Starčević also refused to accept the use of the Cyrillic alphabet.

Josip Frank was a follower of Starčević, and a fanatical opponent of any alliance between Serbs and Croats, as well as unification of Croatia with Serbia. He was also an instigator of the persecution of Serbs on Croatian territory. His followers came to be known as "Frankovtsi," and after the fall of Yugoslavia in 1941 they became supporters of the Axis satellite regime in Croatia. Starčević was anti-Habsburg. Frank was not; he tied the cause of Croatian chauvinism to the black-and-yellow banner of the Dual Monarchy.

In spite of all the difficulties—real or perceived—among the South Slavs there were dreamers who were not to be deterred. These were mainly Croatian and Serbian intellectuals, who struck a responsive chord with idealistic youth. The peasant masses generally did not engage in the romantic emotionalism of the Yugoslav idea. Yet the intellectuals were not operating in a vacuum. There were people-to-people contacts, especially in the border areas, where people crossed boundaries to go to church services, athletic events, and other outings. Serbian singing societies and gymnastic *(sokol)* organizations sent representatives to perform before South Slav audiences in towns and cities that were a part of the Austro-Hungarian Empire. Such mingling encouraged intellectuals to push toward a realization of their idea.

There were several proposals on how to bring about changes in the decaying Ottoman Empire. And inside the Austro-Hungarian Empire there were voices (Croat, Slovene, Serb, Austrian, Hungarian) raised on how the Dual Monarchy should be organized in a more representative fashion. Some saw the solution in triadism—German, Hungarian, and Slav components. Others thought in terms of federalism and centralism. In the years following the failure of the revolution of 1848, Croatian thinkers increasingly turned to the idea that the solution to their self-governance lay outside Austria-Hungary. This signified an increasing role for Serbia and perhaps a South Slav state.

Serbia, for its part, was aware that its recently acquired independence was still fragile, and some of its leaders believed that they should ensure national survival by a mission to unite Serbs in scattered parts of the Habsburg and Ottoman empires. A document known as the Načertanije (Outline), produced in 1844 by the rising Serbian political star Ilija Garasanin, became the most important program for Serbia's foreign policy in the nineteenth century. It said, in part, that Serbia must realize that it is still small, and could achieve its future "only in alliance with other neighboring peoples." Garašanin was aware, of course, that Serbs lived outside Serbia's borders, but how comprehensive his understanding of Croats and Slovenes or South Slavs in Macedonia and elsewhere was, is not known. There is debate among scholars about whether he was thinking in terms of a Great Serbia or, quite radically, of a Yugoslav state, but near the end of his political career (1867), he proposed to Croatian Bishop Strossmayer the unification of Yugoslav *plemena* (tribes or clans) in one united state. Serbia's best hope, he thought, was reliance on France and England. He put no faith in Austrian or Russian policies.

Bishop Juraj Strossmayer was a significant figure. Garašanin first met him in 1852 and was impressed. Garašanin had sent agents into South Slav areas of the Dual Monarchy, and some of these made friends with Croats, among them Franjo Rački, a friend of the bishop, and Matija Ban and Ljudevit Gaj. Meanwhile, Serbs in Austrian territory were also promoting the cause of South Slav unity. The most noted were Svetozar Miletić and Vladimir Jovanović.

The important point is that no one person or group deserves exclusive credit for the Yugoslav idea. Different persons or groups did different things. Matija Ban, for example, founded, in 1860 in Belgrade, a secret committee to prepare a general Balkan uprising. There were various Serbian leagues whose purpose was to bring about a revolt in Turkish-controlled areas. Young Serbs, Croats, Slovenes, and Bulgarians corresponded with the noted Italian nationalist Giuseppe Mazzini, in the hope of gaining support abroad.

Great hopes had been built up in the belief that Garašanin and his prince, Mihailo Obrenović—Karadjordjević and Obrenović were the two Serbian royal lines—would launch a general Balkan uprising against the Turks. This was especially true after the Turks, in 1867, vacated their important fortresses in Serbia. Garašanin's fall from power as prime minister and minister of foreign affairs came in the same year, 1868, as the assassination of the prince, which ended these hopes.

This left the so-called Eastern Question, which was based on two current conditions: the conflict of the great powers, and the aims of the Balkan peoples as they asked what was to happen to the European part of the decaying Ottoman Empire. The answer was not simple.

The forty years preceding World War I saw the great powers frustrating the dreams of the South Slavs. In the course of several crises during that period, Serbia, the main hope of the South Slavs, suffered setbacks. At the same time, Austria-Hungary was gaining control over areas given up by the failing Ottoman Empire. The 1875–78 crises were crucial.

These involved wars between Russia and Turkey, with Serbia and Montenegro as allies of Russia. In the peace of San Stefano, in 1878, Russia betrayed the Serbs and created a larger Bulgaria at their expense. Although Serbia and Montenegro made important gains at the Congress of Berlin, in a subsequent revision of the San Stefano accords, Serbia lost a great deal more when the great powers acquiesced in Austria-Hungary's determination to occupy Bosnia and Herzegovina, earlier a part of the Ottoman Empire. Serbia's leaders had had high expectations that these provinces would be joined to Serbia.

The one ray of hope came from the fact that Bosnia and Herzegovina had not actually been annexed, but merely "occupied." The goal of liberation was not dead, only dormant. What the Serbs did not know, however, was that there was an understanding among the great powers that at some future date Russia would give its prior assent to annexation; this did not occur until 1908, in effect after a Vienna *fait accompli.*

Conceived at about the same time as the annexation, and also made public only in 1908, was Austria-Hungary's plan to build a railroad across the Sanjak of Novi Pazar, a narrow strip of land linking Bosnia-Herzegovina and the Turkish-held province of Kosovo. The annexation plus the railroad would give Vienna a piece of territory that would keep Serbia from gaining access to the Adriatic, to Montenegro, and to Greece. At the same time it would give Austria-Hungary access to Macedonia and the port of Salonika, in Greece, without forcing it to go through Serbia, and also direct access to Bulgaria. In short, Serbia would be encircled.

In the end, this imperial project did not materialize. Even

so, Austria-Hungary felt increasingly threatened by the attraction of its South Slav inhabitants to Serbia. This was especially true after 1903 when in Belgrade the Obrenović dynasty was overthrown. It had been reasonably cooperative with Vienna. Now came the establishment of a firm parliamentary democratic order under a constitutional monarch, Peter I Karadjordjević. Thus Serbia soon became an even stronger magnet for the South Slavs in the Dual Monarchy.

Vienna was determined to weaken, or even, if events proved favorable, destroy Serbia. And it sought to do it soon. To this end, it launched two main courses of action—a tariff war and the outright annexation of Bosnia-Herzegovina.

The tariff war, designed to make Serbia subservient economically to Vienna, lasted from 1906 to 1911. The pretext for closing off Serbian trade was Serbia's determination to act like an independent state, to determine its own foreign and defense policy. Specifically, Serbia refused to seek loans from Austria-Hungary to buy armaments and railroad-building equipment. Instead, it turned to the West, mainly France, and received loans from Paris to buy French artillery for its army. During the reign of the Obrenovićes, Austria-Hungary had not only enjoyed a favorable balance of trade with Serbia, but also monopolized armaments sales to Belgrade.

When Vienna's tariff war against Serbia failed of its objective, the leaders in Vienna realized that additional action against Serbia would be necessary. In October 1908, they announced the annexation of Bosnia-Herzegovina. Russia was caught by surprise; as noted earlier, Russia assumed that the great powers' understanding provided that Russia would be informed well in advance, so that its consent would precede such an act. Nevertheless, Russia was not in a position to prevent the annexation, in good part because it was weakened by its defeat by Japan in 1904 and by revolutionary turmoil at home in 1905. Russia was forced by the other powers to exert pressure on the Serbian government to declare in writing that Serbian rights had not been impaired and to cease its protests.

This was not the end of Austria-Hungary's moves against Serbia. It engineered treason trials in Cetinje, the capital of Montenegro, and in Zagreb, in the hope of discrediting Belgrade. The trial in Cetinje in 1908 was designed to prove that Serbia was involved in a plot against Prince Nikola of Montenegro. In 1909, the trial for treason in Zagreb aimed at proving that the Belgrade cabinet was working for revolutionary action among Serbs and other South Slavs in the Dual Monarchy. The documents produced in Zagreb, and later at a trial in Vienna, were soon proved to be patent forgeries.

The tariff war was sometimes referred to as the "pig" war, because it cut off Serbia's most important export to Vienna, hogs, and it no doubt hurt the Serbs. But the annexation was regarded by the Serbs as possibly a permanent blow to their national interests. There was a substantial public outcry for war, but the Serbian leaders, though outraged, knew that the country was in no position to fight the empire. What should their strategy be?

The old and continuing desire to force the Turks out of Europe was given new strength. Earlier, Serbia had several times sought to improve relations with Bulgaria, which had, for many decades, followed the lead of Russia, but then, disenchanted, had turned to Austria-Hungary. From the middle of the nineteenth century to 1910, Bulgaria was hostile to Serbia, although both were largely Slavic and Orthodox. By 1911, however, the leaders in Sofia realized that war with Turkey—clearly probable at the time—would necessitate an alliance with Serbia. For Serbia and Montenegro, too, alliance with Bulgaria was imperative, because Austria-Hungary, bloated with victory after the annexation of Bosnia-Herzegovina, was moving to encircle the free Serbs in the south and southwest.

Central to a Balkan alliance was the agreement between Serbia and Bulgaria signed in February 1912, and soon adhered to by Greece and Montenegro. The agreement was, however, seriously flawed. None of the Serbian leaders wanted to take any responsibility for the secret annex to it that ceded to Bulgaria what they regarded as Serbian lands in Macedonia.

The secret annex stipulated how Macedonia would be divided in the event of war with Turkey. Two zones were uncontested, but between them was disputed land. The agreement provided that the Russian tsar would serve as arbiter should the Bulgarians and the Serbs fail to agree on boundaries. Top Serbian military men did not like the agreement, but found the urgings of Serbia's prime minister, Nikola Pašić, persuasive. He had intimate knowledge of the problem, since he came from the area bordering on Bulgaria and had spent several years in political exile in that country.

The Turks, sensing danger, mobilized in September, and the Balkan states followed a few days later. The First Balkan War began in October 1912. In November, the Turks sued for peace, asking the great powers to intervene. Their defeat surprised most of Europe. Austro-Hungarian circles had in fact anticipated that Turkey would easily defeat the upstart Balkan nations.

The Serbian army had been particularly effective and had been called upon by the Bulgarians to strengthen their own effort. The inability or unwillingness of the Bulgarians to contribute their share to the conflict became a significant factor in postwar disagreement between Serbs and Bulgars. While Turkey was waiting for the great powers to make peace, Serbian troops drove to the Adriatic and established a common frontier with Montenegro. Austria-Hungary demanded that the Serbs withdraw. Serbia and Montenegro presented their case to the great powers, partly on assurance by the Russians that they would defend Serbian interests. But the great powers, seeking to appease Vienna, decided on the creation of an Albanian, mainly Muslim-populated, state, and Serbia promised to withdraw its troops once a peace treaty was signed.

In the meantime, Montenegrin troops were laying siege to Skadar (Scutari)—the area around the largest lake in the Balkans—and asked for Serbian help, which was forthcoming. In April 1913, Skadar surrendered to King Nikola of Montenegro, but the victory was short-lived. The great powers insisted that Skadar belonged to Albania, and again demanded that Serbia and Montenegro withdraw. Both were thus denied significant victories. Although in a sense it could be said that Austria-Hungary triumphed, its actions were to lead to its destruction.

The problem of dividing the lands liberated from the Turks was complex, particularly in drawing the boundary between Serbia and Bulgaria. The Serbs proposed a revision of their agreement in view of the fact that the Bulgarians had not lived up to their obligations in joint military operations against Turkey. Serbia and Montenegro, resentful toward the great powers for forcing them to give up victories in the west, wanted that to be considered. To add ammunition to this explosive situation, the Bulgarians alienated the Romanians by insisting on retaining the province of Dobrudja, one claimed by the Romanians.

Since Bulgaria was in no mood to negotiate, the Serbs turned to the Russian tsar, as provided in the alliance agreement, even though their past experience had raised serious doubts about Russian neutrality. Bulgaria did not wait for arbitration. In June 1913, it launched a military strike against the Serbs and Greeks, in what became the Second Balkan War. The Romanians joined Greece, Serbia, and Montenegro in winning a quick victory over Sofia's forces. There was no need for Russian intervention. The tsar accepted the outcome.

As a result of the two Balkan wars, Serbia almost doubled its territory and increased its population by nearly fifty percent: from around two million at the time of the Congress of Berlin (1878) to four million in 1913. This was not achieved without heavy cost. Political, economic, and administrative problems were doubly aggravated.

The fact that Serbia was weakened by the great powers' intervention in 1912 did not comfort the rulers in Vienna. Adding to their concern was the formation in 1911 of the Serbian organization Union or Death, which in popular parlance came to be known as the Black Hand. Its openly declared aim was to work for the union of all Serbian lands—in Croatia, Bosnia, and Herzegovina especially—by whatever means necessary.

The Habsburg dynasty was also disturbed by the democratic society established in Serbia. In the decade before World War I, Serbia had a functioning democratic political system that was in every respect on a par with those of contemporary France, the Netherlands, and Denmark. Serbian cabinets were chosen by free and open elections and could always be voted out of office in parliament. Moreover, Serbia was, in the pre-1914 years, a country of balanced budgets. It should be noted that in terms of political democracy, Serbia was ahead of Germany, Russia, and Austria-Hungary, as well as all the other Balkan states. Montenegro could claim some democracy, since a popular referendum had chosen the Njegoši to serve as rulers. Before the mid-nineteenth century, the prince of Montenegro was also its bishop, and succession proceeded from uncle to nephew. Nikola had become prince in 1860 without a clerical role; he was named king in 1910.

The assassination of the heir to the Habsburg throne, Archduke Franz Ferdinand, in June 1914 in Sarajevo, on the soil of the Dual Monarchy by one of its subjects, gave the government in Vienna the pretext to crush Serbia. The assassin was a Serb, and to Austro-Hungarian leaders this was proof that agents from Serbia had directed the assassination with the connivance of officials in Serbia. There is evidence that Serbian intelligence was operating in Bosnia-Herzegovina, but there is no conclusive proof of official Serbian involvement in, or knowledge of, any plot to kill the archduke.

On the other hand, there is the extraordinary circumstance that a Habsburg archduke who was in line to succeed the very old emperor chose to visit Bosnia, where almost all of the population was Slavic, whether Christian or Muslim, on a day revered by all Serbs. Vidovdan ("the Day When We Shall See") occurs on June 28, and to this day is celebrated by Serbs. It is the day of the Battle of Kosovo, when the Serbian Empire met defeat by the Turks in 1389. The Serbian government had been worried that the archduke was to inspect Austrian troops on Serbia's most important national holiday, and had asked its minister in Vienna to caution the government about the selection of that date.

It is important to note that before the assassination, the leaders in Vienna, in asking for the support of their German allies in a preventive war against Serbia, had argued that it was an absolute necessity for their salvation. Berlin accepted the eventuality, but was uneasy about the weakness of the Dual Monarchy. Belgrade, too, was convinced, before the assassination, that an attack on Serbia would come. In the spring of 1914, Prime Minister Pašić took his fears to the tsar in St. Petersburg. The tsar was incredulous, but finally promised Russian protection in case of an unprovoked attack.

In July, nearly a month after the assassination, Vienna presented Belgrade with a forty-eight-hour ultimatum, which was humiliating in all respects—an ultimatum no independent state could accept without giving up its independence. Among other conditions, Austria-Hungary demanded that its own police and army have the right to investigate persons and groups in Serbia with impunity. Serbia was willing, in a conciliatory manner, to accept most of the provisions, but it could not accept the destruction of Serbian sovereignty. Vienna, bent on war, was adamant.

Neither the Austrians nor the Germans had been able to find any evidence of official Serbian complicity in the assassination of the archduke. The most that Vienna could do was demonstrate that the Serbian cabinet had tolerated the existence of anti-Austrian organizations in Serbia.

In this light, Germany was satisfied with the Serbian answer to the ultimatum. Austria-Hungary, however, was not. It declared war on Serbia on July 28, 1914, acting hastily on the assumption that Berlin would approve. Although this was not what Berlin wanted, Berlin proceeded to declare war against Serbia.

The Serbs and Montenegrins resisted the Austro-Hungarian invasion and in the first year of the war drove the invaders from Serbian soil. But when German troops arrived in 1915 with superior armament, and when Bulgaria suddenly attacked Serbia's weakened eastern flank, the Serbian army was forced to withdraw. In the winter of 1915–16, Serbian troops, government, Prince Regent Alexander and his ailing father, King Peter I, together with fleeing civilians, made their way through Montenegro and over the Albanian mountains. Losses were heavy from illness and disease, as well as from the fighting. Some 150,000 troops and civilians at last reached the Greek island of Corfu. There the army recouped, and in 1916 returned to fight, on the Salonika front, and then made its way back to the homeland. With that war's end, the Balkans were never again captive to two old, and now-destroyed, empires, Habsburg and Ottoman. They now had the opportunity to create a free Slavic state or states.

Before proceeding, some words of caution seem desirable. The differences among the South Slavs, noted earlier, have in recent years often been magnified in the popular media outside Yugoslavia. For example, it is often said that the

Croats and Slovenes are "western" and the Serbs "eastern," as if to suggest an unbridgeable gap between them. To be sure, the Croats and Slovenes, especially the latter, were shaped more by Western European Roman Catholic values than were the Serbs, but that hardly means that Eastern Orthodox Christians were without art, literature, music, and philosophy. To suggest that would be to suggest that Greeks, freed from Ottoman rule later than Serbs, are not part of European culture.

At the same time, in writing of differences between Croats and Serbs especially, it must be remembered that religion has been a most divisive force. The flames of hatred were often fanned by extremist Croat Catholic priests. There is no evidence of similar actions on the part of Serbian Orthodox priests or Slavic Muslim religious leaders.

The egregious East-West distinction is not the creation of the present-day press. For example, British historian R. W. Seton-Watson regarded the Dual Monarchy as untenable, but he held some biased views of its fall and those who would succeed to authority. In a book published in 1911, he wrote: "The triumph of the Pan-Serb idea would mean the triumph of Eastern over Western culture, and would be a fatal blow to progress and modern development throughout the Balkans."

Modern observers who believe that the Serbs are politically shaped by the "backward East" need to be reminded that all those who in the latter half of the nineteenth century and the beginning of the twentieth directed Serbia's development toward freedom and democracy—notably, Jovan Ristić, Prince Mihailo Obrenović, Prime Minister Nikola Pašić, and King Peter I Karadjordjević—spent a great deal of time in the West or at Western European universities. They were all imbued by a mature appreciation of Western democratic ideas. King Peter I, who was a constitutional monarch in the best sense of that term, had, before coming to the throne, translated John Stuart Mill's essay "On Liberty." Serbian intellectuals had translated works of other Western democratic thinkers. The prince-bishop of Montenegro, Peter II (1812–1851), the greatest Serbian poet of all time, had read John Milton, studied classical philosophy, and visited Vienna and St. Petersburg.

The South Slavs are ethnically alike, and the Serbs and Croats speak the same language, as has been said. The one may call himself Croatian, and the other, Serbian, but the language is Serbo-Croatian, or Croato-Serbian. There are a few differences, but these are not any greater than the differences between American English and British English. The Slovenes speak a Slavic tongue similar to Serbo-Croatian, and they have little difficulty in understanding Croats or Serbs. The Macedonians developed a separate written language after World War II, but it derives in part from Serbo-Croatian and Bulgarian, which is also a Slavic language. Macedonia, like Serbia, is Orthodox Christian, although Tito did insist on creating a Macedonian Church, separate from the Serbian one.

In brief, the things that the South Slavs have in common logically should outweigh the things that divide them.

2

The Creation of the Yugoslav State

The most important prerequisite for the realization of the South Slavs' dream of uniting within one state was the downfall of two empires—the Ottoman and the Habsburg. At the outbreak of World War I, the former was in the process of passing into history, but the latter was diplomatically active and belligerent, and little sign of decay could be seen. Moreover, the Allies—Great Britain, France, and Russia—had not made the destruction of the Austro-Hungarian Empire one of their war aims. Hence the South Slavs, especially Serbia and Montenegro, the small allies that were to become central in the realization of a single Yugoslav state, confronted an immense obstacle: They were allied with great powers, yet their fellow warriors were not committed to the dissolution of their common enemy, which held captive Poles, Czechs, Slovaks, Slovenes, Croats, and many Serbs.

The Allied great powers favored saving the Habsburg monarchy for different reasons. Britain did not wish to see Serbia expand, because it looked upon Serbia as Russia's natural ally, a means of spreading Russian influence into the Balkans. France viewed Germany as the main enemy, not Austria-Hungary. Russia was opposed to dismantling the

Habsburg monarchy for dynastic reasons, but was also concerned that Orthodoxy would suffer because the Catholic Slovenes and Croats could spread Vatican influence in a Yugoslav state. In addition, the Entente powers had seen no concrete indications that the Croats and Slovenes really wanted to unite with Serbia. Most important, however, was their desperate desire in 1914 and early 1915 to get Italy, Bulgaria, and Romania into the war on their side. This would require concessions to Italy of parts of the eastern Adriatic coastline, to Bulgaria of land in Serbian Macedonia, and to Romania of territory in the Banat, part of Serbia. Though some concessions would be at the expense of the Serbs, those given to Italy would be mostly at the expense of Croats and Slovenes.

Any argument that Serbs primarily sought a Great Serbia is weakened by their refusal in 1915 to be appeased by the Allied powers' handing out prizes to Bulgaria and Romania and Italy. The Entente powers had offered to compensate their ally Serbia by promising that after the war Serbia would get Bosnia-Herzegovina and other areas of Austria-Hungary populated by Serbs, as well as a large part of Dalmatia and part of Albania. The Serbian government agreed reluctantly to some of the promises made to Bulgaria, but would not accept the Entente plan to create a much larger Serbia geographically. In any case, the Serbian leaders had already begun work on their own larger goal—creating a South Slav state—and hoped to persuade their allies to support that goal.

Several months before Vienna's ultimatum to Serbia, Pašić had visited Tsar Nicholas in search of a suitable princess for the Serbian throne. On that occasion he had observed that, if circumstances permitted, she might become princess of the "Serbo-Croatian Yugoslav people." After the war had begun, Prince Regent Alexander—son of the ailing King Peter I—in his order to the army, mentioned the millions of our brothers "from Bosnia-Herzegovina, Banat and Bačka, Croatia, Slovenia, Srem, and Dalmatia."

Pašić, within a month of the outbreak of hostilities in 1914, had the Ministry of Foreign Affairs begin preparing formulations and explanations of the Yugoslav problem. He assembled a group of outstanding ethnologists, geographers, historians, economists, and international lawyers to prepare studies that would justify the goal of a common South Slav state. In September, he sent instructions to Serbia's representatives in allied capitals, setting forth in fairly precise form the projected territorial boundaries of a new Yugoslav state after the defeat of Austria-Hungary.

The war provided the possibility, for the first time, to realize the South Slavs' dream of unity. Yet when it came, they were utterly unprepared. They had not defined, in practical terms, how to proceed with the formation of a new state. No studies were ready. No nucleus of leaders had come together. Although some leaders had no doubt thought in practical terms, no concrete steps had been taken.

There was a historical solution to this situation, which popular accounts of the present struggle between Croats and Serbs have missed. During the nineteenth century, in the unification of Germany and Italy, a dominant state, among many small principalities, took the lead. In Germany it was Prussia. In Italy it was Piedmont. If the Croats, Slovenes, and Serbs were to form a South Slav state, it was not surprising, therefore, that Serbs took the lead. They had had long and recent experience with independence, in both Serbia and Montenegro.

The Serbian Prime Minister, Pašić, in a move that many viewed as pure bravado, in December 1914 delivered the first public declaration of Serbia's war aims. The cabinet quickly approved, and it came to be known as the Niš Declaration (from the Serbian city where the government met after the fall of Belgrade). The Serbian parliament then gave its approval. The declaration stated that Serbia's main intention, after victory, was liberation and unification of all Serbs, Croats, and Slovenes. Serbia's great-power allies did not accept the Niš Declaration, but the Serbian parliament, in a secret session about a year later, renewed its support.

Pašić defended the declaration in terms of maintaining peace in the Balkans, which could most effectively be done by creating "one national state, geographically sufficiently large, ethnographically compact, politically strong, economically independent, and in harmony with European culture and progress."

The problem for the Serbian leaders was to prevent this war aim from being ignored, especially by the Allies. This became more important and more difficult when superior German armament forced the Serbian government and army to retreat across the Albanian mountains to Corfu.

Even before the declaration, Pašić had recognized the need for propaganda work in the main allied nations. It was necessary, he believed, to have an organization to assure that press accounts continued to appear about Serbia—its brave army, its democratic society, its enormous sacrifices in lives, and its aspirations. In pursuit of this goal, he sent two Serbs from Bosnia to Italy in November 1914 to work out a common front with Croatian and other South Slav exiles from Austria-Hungary. The main Croatian leaders were Franjo Supilo, Hinko Hinković, Ivan Meštrović, and Ante Trumbić. With other South Slavs, they formed the Yugoslav Committee; Trumbić was chairman. The role of the committee initially was to inform the Allies concerning the plight of the South Slavs in Austria-Hungary and their desire to unite with Serbia to form a South Slav state, which would necessitate the destruction of the Dual Monarchy. When the Serbian government learned that, to persuade Italy to join the Allies, Britain, France, and Russia had in 1915 signed a secret treaty in London granting important concessions to Italy at the expense of lands inhabited by South Slavs, Pašić asked the committee to work energetically to nullify that treaty.

Although Pašić needed the committee, the "secret" treaty made the committee need Pašić and Serbia far more urgently. Earlier, some of its members had seen themselves as spokesmen for their peoples in Austria-Hungary, even those who dreamed of forming an independent state apart from Serbia. But when they learned of the treaty, they saw that any such dream was doomed. Their only chance for independence was to work in concert with Serbia for the destruction of the Dual Monarchy.

At the same time, the committee's Croatian members decided they were not equal partners. They felt like outsiders, because they were never as well informed as Pašić and his cabinet colleagues, who met regularly with representatives of the Allied governments, and who then passed on to the committee such information as they chose. Consequently, some committee members, notably Trumbić, sought information independently and separately made efforts to influence the Allies. Their best contacts were two Englishmen, Wickham Steed and R. W. Seton-Watson, who were not known to feel friendly toward Pašić and Serbia. These committee members found a sympathetic response also among the minority-party members of Pašić's coalition government.

Pašić was aware that the committee's Croats were emboldened by Serbia's weak position and had begun agitating for a more decisive voice in shaping the future. He considered this divisive and felt that they made more difficult his work with the Allies to rehabilitate the remnants of Serbia's army so that it could return to battle. He could not ignore the fact that the committee, in presenting memorandums to Allied governments, was purporting to be the representative of all South Slavs under the Habsburg monarchy, and even those in North and South America. The committee was, in effect, asking to share power with the Serbian government. Pašić, of course, could not constitutionally share his or his cabinet's power. Nor could he speak for another Serb Allied combatant, Montenegro. But he was convinced that in dealing with the Allies, the South Slavs would be most effective if they spoke with one voice.

At the same time, he faced three problems: first, the Treaty of London—which, if adhered to after the war, would make the formation of a Yugoslav state exceedingly difficult, if not impossible; second, Russia, Pašić's "ace in the hole," which could no longer be counted on after the revolution in March 1917 changed relationships and alliances; third, the fact that in May South Slav deputies in the Austro-Hungarian parliament had issued a declaration urging the union of the provinces where Slovenes, Croats, and Serbs lived in one state "under the scepter of the Habsburg dynasty."

Under the circumstances, Pašić was left no choice but to seek an understanding with the Yugoslav Committee. In July 1917, they met on Corfu, and out of the meetings came the Corfu Declaration, signed on July 20. Full of sentiments of unity, brotherhood, and common interests, the declaration consisted of fourteen specific points concerning the organization of their common future state.

Since the great powers were still unwilling to consider or acknowledge that such a future state was one of their war aims, the declaration was in part designed to exert pressure on them to accept the consequences of the dissolution of the Habsburg Empire. It was also felt to be an answer to the May declaration of the South Slav deputies in the Austro-Hungarian parliament.

The principal negotiators at Corfu were Pašić, on behalf of the Serbian government, and Trumbić, on behalf of the Yugoslav Committee. The first point of the declaration says that the State of the Serbs, Croats, and Slovenes (its official name) will be a "constitutional, democratic, and parliamentary monarchy with the Karadjordjević dynasty at the head." Aside from the dynasty, Serbia did not ask for privileged status or veto power, as had Prussia, for example, when it was the central force in unifying Germany. Moreover, Serbia was willing to give up its democratic constitution, convinced that a constituent assembly would produce a constitution that would be as good. In addition, Serbia gave up its flag, coat of arms, and other national symbols. Most important was Serbia's willingness to present the Corfu Declaration to the Allies as a program for the liberation and unification of all Serbs, Croats, and Slovenes. This was a blow not only to Austria-Hungary, but also to Italian aspirations in the northern Adriatic.

Pašić and his cabinet colleagues, as well as a majority of the Yugoslav Committee, agreed that the future state should be a unitary one, with certain local self-government rights. The unitary principle of governmental organization followed the practice of all other European states except Switzerland. The final point of the declaration stated that a constituent assembly would adopt a constitution "by a numerically qualified majority," a somewhat vague and imprecise provision that was nowhere defined.

The members of the Yugoslav Committee had had to face compromise. Agreement on certain principles—for example, parliamentary democracy—was not difficult. Some committee members would have preferred a republic, but they were in no position to oppose the monarchy or its right to exercise the powers of a constitutional monarchy. Serbia's monarchs had ruled since 1804 and were no mere symbol, but, rather, instruments in Serbia's fight to free itself from the Turks.

The committee, and especially its Croatian members, wanted a requirement that the future constitution needed to be ratified by a majority of each national group or by a two-thirds majority of the constituent assembly. Such a question is present in every democracy and is answered in various ways. There was no accepted Western "blueprint" in 1917, or even now. They had to settle for the words "a numerically qualified majority." The constituent assembly, when it met in 1921, concluded that an absolute majority of its delegates was sufficient to meet the language of Corfu.

Ironically, relations between the committee and Pašić worsened after Corfu. Following the signing of the declaration, Trumbić paid high praise to Serbia: "As a state she has made the greatest sacrifice for the union of our three-named people. She announces that she is ready to sacrifice her state individuality in order that one common state of all Serbs, Croats, and Slovenes be created. With that she begins the greatest of her works and attains the absolute right to be called the Yugoslav Piedmont." (There was another Serb sacrifice, which was little noted. Montenegro, a free state older than the U.S. republic, was to give up its independence and monarchy to become part of the new union.)

By the end of 1917, there were storm signals on the horizon. Pašić's position, already weak, was made more difficult by the Bolshevik seizure of power in Russia in November 1917. His ally Russia was soon out of the war. The committee sought to exploit Pašić's weakness, convinced that it could get more than the Corfu Declaration offered. Members were encouraged in 1918 by the arrival in Europe of more and more American troops, with the promise of an Allied victory. They therefore increasingly attempted to get for themselves more of a decision-making role in the Serbian cabinet.

Yet there was a vacuum in which mind-changing could float somewhat freely, because both Pašić and the committee were aware that as of January 1918 the Western Allies had still not indicated a desire to see the end of the Habsburg monarchy. Earlier, Austria-Hungary had indicated a willingness to sign a separate peace, and the Western Allies had been receptive. Who was to say what compromise might be reached? President Wilson and Prime Minister Lloyd George had said that they favored autonomy for the oppressed nationalities in the Habsburg monarchy, but there was no sign in their statements that they would support a breakup of the Habsburg state. If that state survived, it would make it easier for the Allies to live up to the promises of land made to Italy.

In these circumstances, Pašić, in the latter half of January 1918, sought to determine the intentions of the Allies. He noted that in the statements of Lloyd George and Wilson there were demands for the liberation of Italians and Romanians, but not for the liberation of the South Slavs. In addition, their pronouncements called for revision of the treaty that gave Alsace-Lorraine to Germany, but did not ask for the revision of the Treaty of Berlin, which had in effect turned Bosnia-Herzegovina over to Austria-Hungary. He asked his ministers in London and Washington to find out if these incongruities were accidental or intentional.

When news of his directive became known to the Yugoslav Committee, certain members interpreted his action as, perhaps, a willingness to abandon the Yugoslav solution to the South Slav question and to settle for territorial acquisitions promised to Serbia earlier. "Great Serbia" was always available as a weapon in the war of words. So Pašić's efforts to ascertain Allied intentions served only to increase tensions between him and committee leaders. Stung by charges that Serbia was imperialistic and that it wanted to dominate in the new state, Pašić in mid-October 1918 gave statements to London's *Times,* the *Morning Post,* and the *Manchester Guardian* stating that "the Serbian people cannot wish a dominant position in the future kingdom of the Serbs, Croats, and Slovenes." At the same time, he asserted that Serbia considered it to be its national duty "to liberate all Serbs, Croats, and Slovenes." When they are free, he said, "they will be guaranteed the right of self-determination, i.e., the right freely to declare whether they wish to unite with Serbia on the basis of the Corfu Declaration, or to create small states as in the distant past," and added that "we will not invoke the Corfu Declaration if that would not correspond to their wishes."

Trumbić and others leveled further attacks on Pašić, contending that the statements to the British press constituted additional proof that he really wanted a Great Serbia instead of a Yugoslav state.

Pašić defended himself by noting that some Englishmen (Steed, Seton-Watson, and others) had told him that he was making unification difficult, that he wanted to put everything under Serbia, that he was bent on annexation and rule by force. In view of such sentiments, he said, he could not avoid saying that Serbia did not want to force Croats and Slovenes into a union. He wanted all Serbs to unite. If others wanted to join voluntarily, well and good, but if not, let them remain where they are. This, of course, was the last thing that Trumbić and his friends wanted to hear, because it would mean that the Croats and Slovenes would remain under foreign rule, or, at best, be left to create independent states on exceedingly small pieces of territory. The parallel with 1990–92 is obvious. Croatia, consisting only of lands with Croat majorities, is not a large area, even though Tito gave Croatia Serb-occupied lands after World War II.

When, in June 1918, the Allied governments recognized the right to independence of the Polish and Czech peoples, the Yugoslav Committee openly demanded that it be internationally recognized as the representative of the Austro-Hungarian Yugoslavs. Pašić was opposed, because he feared that the Allies might play one group of South Slavs against another. He did offer to include in his cabinet three Austro-Hungarian South Slav representatives, but Trumbić refused.

By October 1918, the tempo of Yugoslav developments had increased sharply, tending to overtake all South Slav leaders. In a desperate effort on October 15, Pašić begged British Prime Minister Lloyd George to support the creation of a Yugoslav state as envisioned by the Corfu Declaration. Lloyd George's response was not reassuring. He told Pašić that if the war continued, and Serbian troops moved into the South Slav areas of the Habsburg monarchy, "your wishes will be fulfilled," but if the war ends, "discussion must take place." Emphasizing that the main thing was for the blood-letting to stop, Lloyd George concluded that the question "is, will your army enter those regions or will it come to discussions."

Meanwhile, the committee was determined to impose conditions on Pašić in advance of the anticipated creation of a Yugoslav state. An ill-concealed campaign to undermine him was building, a campaign that enlisted British friends. At the same time, Austro-Hungarian forces were collapsing, and the picture changed. The most important event was the announcement in Zagreb on October 29 of the formation of a new state of Slovenes, Croats, and Serbs, and the creation of the National Council as its governing body.

The establishment of the National Council was at once welcomed by the Yugoslav Committee and yet regretted, because the National Council seemed about to replace the committee as the voice of the South Slavs within the disin-

tegrating Habsburg monarchy. Rather than be left out entirely, Trumbić and his friends sought international
recognition for the council, but at the same time seized the
propitious moment to try to extract concessions from Pašić.
Serbia's prime minister, seeking to minimize the impact of
the discordant voices, asked Prince Regent Alexander's permission to undertake talks that could lead to a coalition
cabinet.

When Pašić arrived in Geneva, the site of the talks, he
found himself engaged in much more widespread and un-
structured discussions than he had anticipated. Critical for
him was the fact that he alone represented Serbia, whereas
he faced three representatives from the National Council,
three from the Yugoslav Committee, and three from his own
political opposition party in Serbia. Outnumbered nine to
one, he decided that in the interests of national unity he
should seek an agreement, even though he knew it would
not be legally binding.

An agreement was reached that provided, pending the
meeting of the constituent assembly, for the Serbian government and the National Council to constitute a dual authority, the former in Serbia proper, the latter in the newly formed
Yugoslav state outside Serbia. In addition, a joint cabinet
would be set up in Paris, with six members named by the
Serbian government and six by the National Council. This
agreement, dubbed the Geneva Declaration, was destined to
remain a dead letter. It was rejected by the Serbian cabinet
sitting in Corfu. By then the Serbian army, which had been
reconstituted and had moved northward from the Salonika
Front in 1916, was on Serbian soil. The agreement was also

rejected by the National Council in Zagreb, even before it heard about the similar action by the Serbian cabinet.

South Slav politicians were overtaken by other events. Before further negotiations between the Serbian government and the National Council could take place, some of the lands for which the council presumed to speak were deciding in favor of union with Serbia. These were Vojvodina, Montenegro, and forty-two out of fifty-two districts of Bosnia-Herzegovina. Within the Slav-inhabited regions of the disintegrating Habsburg monarchy, demobilized Austro-Hungarian veterans, many of whom were armed, angry, and resentful of any state authority, were contributing to a proto-revolutionary situation. Still other areas of what became Yugoslavia were being occupied by Italian troops seeking to assure to Italy the promises of the Treaty of London.

In the light of these developments, representatives of the National Council rushed in the last days of November to liberated Belgrade in an effort to reach agreement with the Serbian government. Prior to sending these delegates, the council, on November 14, 1918, had instructed them to be guided by a number of conditions in connection with the nature and organization of the new state. Among these were stipulations that the constituent assembly decide whether the state should be a republic or a monarchy, that the future constitution be adopted by a two-thirds vote, that only certain specified government functions be lodged in the central government, with remaining ones to be exercised by local units. These stipulations were contrary to the Corfu Declaration. Furthermore, they were in some manner sub rosa, for it is significant that no mention was made of them in the representatives' address to Prince Regent Alexander requesting unification.

In their audience with Alexander, they made the following points: sovereign authority shall be exercised by Alexander; pending convocation of the constituent assembly, an agreement shall be reached on the establishment of a responsible cabinet and a temporary parliament; during the transitional period, each unit shall retain its existing authority, although under the control of the cabinet; and the constituent assembly shall be elected on the basis of direct, universal, equal, and proportional suffrage. No other conditions were advanced.

On December 1, 1918, Alexander accepted their statement and proclaimed the creation of the Kingdom of the Serbs, Croats, and Slovenes.

The implementation of this decision now proceeded. Following the establishment of a provisional parliament, the National Council in Zagreb, on December 28, 1918, and the Serbian Skupština in Belgrade, on December 29, 1918, dissolved themselves, declaring their respective missions accomplished. The long and thorny road leading to the promised land of one South Slav state was at last traversed.

At Versailles, the actions of the South Slavs in establishing the new kingdom were, in effect, ratified, and this occurred in no small measure by virtue of Wilson's assertion that he had not heard of the Treaty of London and would not be bound by it. In Lloyd George's memoirs of the peace conference, he referred to Pašić as one of the "craftiest and most tenacious statesmen of Southeastern Europe. . . . The foundation of the Yugoslav Kingdom was largely his doing.

. . . He took care that this extended realm was an accomplished fact before the Peace Conference had time to approach the problem of adjusting boundaries."

The United States was the first great power to recognize the new state, in February 1919.

3

The Struggle
to Create a Viable
Political System

They had never lived together in one dwelling. They lived in diverse homes, often not their own, and under masters who had different codes of behavior. Although, in essence, they spoke the same language, they knew each other only distantly and incompletely. Most of them were adherents of Christianity, either Orthodox or Roman Catholic, but some of them were believers in Islam. They had often heard romanticized versions of how they were alike and why they should be together, but this had seemed a distant dream. Then suddenly, while their masters were warring, they found themselves free to come together. In an atmosphere of mixed emotions, and with many idealistic expectations, they seized the moment and consummated a hasty union.

Such was the situation of the South Slavs near the end of 1918, when they created the new state known as the Kingdom of the Serbs, Croats, and Slovenes. But who was to be the master in the new household? How were they to assign powers, duties, responsibilities? The sketchy agreements of their quickly assembled leaders provided that Serbia's Prince Regent Alexander would be the single solid symbol of state power as they embarked on the task of nation-building. His actual powers were limited, since he was a constitutional monarch. Real political power would be vested in popularly elected representatives. How were these to be chosen, and how were their spheres of authority to be defined? For the time being, there was at work an interim authority to accomplish the transitional tasks of getting the state's boundaries recognized, signing the peace treaty, and preparing the way for a constitutional convention that would write a constitution for the new state.

The interim authority was primarily a cabinet and a provisional parliament (called the Temporary National Representation), with the cabinet responsible to the parliament, both hastily put together. Some members of parliament had a great deal of political experience; others had little or none at all. Those with considerable experience were mainly Serbs who had held high positions in Serbia's or Montenegro's governments. Others had gained some experience in the Austro-Hungarian parliament, which, however, lacked the powers usually associated with a democratic legislature. Slovenes and Croats who favored the South Slav union brought with them the habits and learned reactions and perceptions that fit their role in an empire dominated by non-Slav masters. Their obstructionist tactics were not venal, generally, but practical in the pre-1914 scheme of things. Thus historical conditioning—experience in a free state and the contrary—influenced the behavior of members of the provisional parliament.

All the political-party leaders recommended to Alexander that he appoint Nikola Pašić as the first prime minister of the new state, but Alexander, perhaps because of disagreements he had had with Pašić during the war, ignored their wishes, justifying his decision by saying that he needed Pašić's skills to head the nation's delegation to the Paris Peace Conference. He chose instead one of Pašić's close collaborators in the Serbian Radical Party, Stojan Protić, to be prime minister.

The tone and style of rule in day-to-day administrative and political matters in the first year was set by the first minister of the interior, Svetozar Pribićević, a Serb member of the Croat-Serb coalition in the former Zagreb Diet, a local government council permitted by the Hungarians. He issued administrative regulations and appointed and dismissed local officials. As a former political co-worker with Croats, he claimed to know them well and was confident he could work with them.

The cabinet—there were to be seven over a thirty-month period, headed by four different men, all Serbs from Serbia— had three broad tasks: to guide the provisional parliament in producing a constitution; to deal with the day-to-day problems of governing until the constitution was adopted; and to attempt to deal with the criticisms and complaints voiced in the temporary legislature. The three were interrelated and complex.

First, however, the cabinet had to organize governmental authority to handle rapidly deteriorating conditions. Before unification the inhabitants had lived under six different legal systems. A special ministry was therefore early given the task of bringing about equalization of the laws and a modicum of legal uniformity. Progress was painfully slow.

In Serbia and Montenegro, enemy occupation had left a wasteland, causing severe shortages of food, clothing, and housing. Yet relatively well-off regions formerly under Austria-Hungary believed they were being drained unfairly to sustain the new state. Some peasants even refused to send food to the cities. Moreover, there was interference with tax collecting and with the drafting of men for the unified army. In addition, rebel bands, including many deserters from military units, roamed the countryside, robbing public offices and destroying utilities.

Cabinet efforts to unify different taxation systems and to set appropriate exchange values for four distinct currencies brought charges of inequality and unfairness. These reactions were relatively minor compared to the armed clashes, disruption of communications, and burning of records in Croatia, which followed upon the government's decree on registering draft animals. The order on branding and recording of livestock was similar to one existing in Serbia before the war, but the idea was foreign to the Croats. The violence that greeted the order was due in part to rumors that the government would confiscate the peasants' livestock.

Although Croatia gave the cabinet its most serious headaches, there were problems of a general nature throughout the country. Initially, aspirations and expectations among the people were high. Recovery, however, proved slow, the effects of the war continued, efforts to establish a workable administrative system were halting, and speculators were seen to be prospering in the midst of human misery. In these circumstances, disappointment and dissatisfaction took root as well-intentioned politicians fell far short of accomplishing what was needed.

And there were foreign threats and problems. Even before the new state could receive international recognition, Italian troops poured into South Slav areas to claim the promises of the Treaty of London. Also before unification, the National Council in Zagreb had asked Serbian authorities to send armed forces to repel the Italians. The Serbian army's presence was both welcomed and resented—welcomed because it was needed, resented because it was not native. A frontier with Italy was agreed to finally, in the Treaty of Rapallo in 1920, but Mussolini continued to create turmoil in the Adriatic during most of the two following decades.

It is not surprising that, given the excessive partisanship and sometimes confused arguments in the provisional parliament, all the cabinets were largely impotent, made up as they were of disparate elements. Complaints about the actions or inactions of government officials were continually voiced. Questions about the future organization of the state were raised but could not be answered.

Highly significant, but little noticed at the time, was the action of Stjepan Radić's small Croatian People's Peasant Party. It decided not to send the few delegates allotted to it to the provisional parliament, choosing instead to attempt to mobilize support for the creation of a Croatian republic. And the few Communists in the South Slav lands were generally scornful of the whole one-state enterprise, asserting that the "real constitution" would be written by "blood in the streets."

So the unraveling of the uneasy consensus had begun before the provisional parliament had met. In addition to Radić's demands for a "neutral peasant Croatian republic" on the eve of unification, the small Croatian Party of (State) Right began a movement on the day after unification to nullify the Act of Union.

As a Croat leader who was not a peasant himself but who knew the Croatian peasantry and managed to identify with it, Radić was adept at articulating its aspirations, aggregating its discontents, and harnessing its national consciousness to his political wagon. Contrary to statements of some scholars in the West, he did not favor a federated Yugoslavia in which Serbia would participate. This is evident from his private correspondence and his public espousal of a Croatian republic. He sent messages to President Wilson and the heads of other states, asking for help in the realization and recognition of a Croatian republic. He wrote letters and pamphlets in an effort to enlist the support of the foreign press.

Initially, Radić was not taken too seriously by the new government. In March of 1919, however, he was sentenced to jail for a year because of his activities. Prison did not seem to inhibit him. Through supporters at home and abroad he continued his separatist pursuits. After his release, he resumed his previous activities and was again imprisoned, but again released—on the day of the elections for the constituent assembly, November 28, 1920.

While in prison, Radić wrote a new Habsburg hymn for what he envisioned as a new country with more than a hundred million people, a commonwealth consisting of ten units—Croat, Slovene, Czech, Slovak, Bulgarian, Romanian, Hungarian, German (Austrian), Polish, and Ukrainian. He envisioned that they would all be Roman Catholic, despite the fact that the Romanians and Bulgarians were almost wholly Orthodox and many Czechs were Protestant.

The elections demonstrated the extent to which the uneasy consensus had come apart in Croatia. Radić's party, once a small, insignificant group, captured nearly all of the seats for Croatian delegates to the constituent assembly, decisively defeating the parties that had represented Croatia in the provisional parliament, parties that had accepted unification under the Karadjordjević monarchy.

Radić interpreted the result as a mandate to create a separate Croatian state. At a mass rally of his supporters soon after, he changed the name of his party to Croatian Republican Peasant Party, and announced that the party would not participate in the deliberations of the constituent assembly. In early 1921, he sent a letter to King Alexander, complaining about his ministers and declaring null and void the request of the National Council for union. He demanded a neutral peasant republic for Croatia, with a separate constituent assembly.

Radić contributed enormously to the new state's insecurity. It had just been formed, and he was trying to tear it apart. He had unified the Croats, whereas the Serbs, who had contributed so much to unification, were divided, leaving the provincial authorities helpless in dealing with him. His seeking foreign backing made things awkward, especially when the Serbs' most experienced leader, Pašić, was at Versailles, heading the Yugoslav delegation to the Peace Conference.

Division among the Serbs was worsened by the formation of a new party, the Democratic Party, as a competitor of the Serbian Radical Party, which had commanded almost universal support among Serbs before World War I. This move did not, however, represent basic disagreement concerning the nature of the state organization. There was impatience and dissatisfaction elsewhere, though, notably in

Macedonia (then known as South Serbia) and in Montenegro, which had surrendered its independence and its Njegoši king to the new state.

Another factor contributing to uncertainty was the electoral law, enacted by the provisional government, which embraced the principle of proportional representation, assuring all groups of participation on the basis of their strength at the polls. As a result, smaller parties benefited at the expense of larger ones.

The elections held on November 28, 1920, were free, secret, and direct, so the delegates to the Constituent Assembly were fairly and honestly chosen. But at the same time, no political party won a majority. Twenty-two political parties or groups participated, and sixteen elected deputies. Of 419 seats, the Democratic Party won 92; the Radical Party, 91; the Communist Party, 59; the Croatian Republican Peasant Party, 50; the Agrarian Party, 39; the Slovene People's Party, 27; and the Yugoslav Muslim Organization, 24. The remaining seats went to smaller parties or groups.

Two unforeseen consequences of these elections were to complicate the task of writing a new constitution. First, the Croatian Union, which until then had represented the Croats in the provisional parliament, was almost annihilated by Radić's party. Second, and more serious, was Radić's decision to boycott the constituent assembly and thus deny fifty elected delegates the opportunity to participate in constitution-making. This act was accompanied by the proclamation of a Croatian republic at a mass rally in Zagreb. It is interesting to note that Radić did not lay claim to certain land areas that the new Croatia in 1991–92 claimed as its own.

The first order of business for the constituent assembly

was to agree on proposed rules of procedure. Some delegates objected to Article 8, which required taking an oath to the king, saying that the gathering should have the power to decide whether the new nation should be a monarchy or a republic. This view was clearly naive in view of the agreement at Corfu and the Act of Union. Interestingly, the Communist delegates, who argued against the oath, finally swore allegiance to the monarch, no doubt preferring a larger field on which to play later.

Another controversial item was the proviso that the constitution could be adopted only if an absolute majority of the total membership voted for it. Some members argued that it should be two-thirds, pointing to the Corfu Declaration's "numerically qualified majority," but those who had voted for an "absolute majority" insisted that they had been faithful to Corfu.

The actual writing of the constitution was guided by Nikola Pašić, who now was made prime minister. Although not fond of him, the king became convinced that his experience and reputation were needed if the kingdom was to get a constitution. Pašić had the support, too, of Svetozar Pribićević, who in the absence of Radić was perhaps the most influential voice from the South Slav areas of the former Austro-Hungarian monarchy.

A number of drafts were presented and debated, but, as might have been expected, the Pašić draft became the working model. The resulting constitution was similar to Serbia's of 1903, which provided for a unitary state and a parliamentary system. Although some consideration was given to a federal organization, Pašić and his collaborators were fearful that federalism for so disparate a group of peoples would lead to disunity and weaken the state internationally. Clearly, it would contribute to multiple political parties based on regional loyalties, which would make agreement on national goals very difficult. Moreover, Pašić was convinced that drawing boundaries for a federation would be so controversial that in the end many Serbs would be left outside the Serbian unit. For these Serbs, who had struggled so long to be united, it would be a disservice, he believed.

The Serbs, in viewing federalism with suspicion, were influenced to no small extent by the example of France, which was once badly divided, but subsequently united into a highly centralized system, particularly in the wake of the French Revolution and the Napoleonic experience.

The Pašić draft, as amended, was adopted on its first test by a vote of 227 to 93. The final version was adopted on June 28, 1921, by a vote of 223 to 35, with 111 abstentions. The chief supporters were the Democrats and the Radicals, but their votes alone would not have been sufficient. Pašić was able to get the votes of the Muslim Organization and some smaller groups. Since 210 was an absolute majority of the total deputies elected (419), the constitution came into force on June 28, the national holiday of all Serbs, Vidovdan, and thereafter came to be called the Vidovdan Constitution.

The cabinet was not pleased with all its aspects, but was convinced that it was important for the new state to have a constitution as soon as possible. It provided for a democratically elected unicameral parliament (Skupština), with ministers responsible both to it and to the king. Members of the Communist Party could not be members, because the party was banned in 1921 for terrorist acts. Civil servants, except ministers and university professors, were also excluded. Military officers and soldiers on active duty did not have the right to vote.

In the spirit of Corfu, the constitution spoke throughout of the "three-named people," the "Serbo-Croatian-Slovenian language," and "Serbo-Croatian-Slovenian nationality." The organization of state authority was made independent of any national or religious considerations. Civil rights—freedom of speech, press, and association—were the same for all citizens, including non-Slavs, such as Hungarians. The judiciary was made independent, and judges were given permanent tenure. Pending necessary legislation, the existing judicial structures were retained.

Thus the new Yugoslav state was launched, but under less than auspicious circumstances. It is fair to say that the Serbs who dominated the constituent assembly did not adequately know Croatia's past, that they insufficiently understood Croatian national consciousness. Yet they had been led to believe one thing by Croatian cooperation before and during the war, and now Stjepan Radić was telling them the opposite. They knew full well that the idea of creating a Croatian republic within the Yugoslav monarchy was a political mirage.

Did they in fact think that Radić would change and that there was hope a consensus could be rebuilt? Some did. Some did not. But the Serbs and others were determined to give the new governmental system an opportunity to prove itself.

Implementing the constitution presented many problems. The Pašić-Pribićević coalition that had framed the constitution immediately undertook several other tasks. Through the minister of the interior it appointed officials and issued instructions on obeying the laws, which was especially critical in Croatia, where obstructionism and resistance were much in evidence. It sought to ensure the integrity of the country's boundaries, threatened by the presence of Italian troops and, in the postwar confusion, moves for a Habsburg restoration. It provided relief to war-devastated areas and inaugurated an orderly demobilization of soldiers, some of whom, Serbs from Montenegro and Serbia, had been in uniform since the Balkan wars of 1912 and 1913. It asked for legislation to adjust and harmonize legal and administrative practices and to staff the courts. Many other urgent problems received only benign neglect, which became a source of considerable discontent.

Complicating the tasks of the first and all subsequent cabinets was the fact that each had to perform as a coalition, because the multiparty system resulted in no political party's having a majority. Moreover, although the Croatian Republican Peasant Party contested elections, it refused to take elected seats in parliament. And the Communist Party was outlawed, under a security statute for the protection of the state enacted after an attempt on the life of Alexander by a Communist in June 1921 and the assassination of the minister of the interior, Milorad Drašković, by another Communist in July.

The first victim of the failure to have a one-party majority in parliament was the Radical-Democratic cabinet, which fell apart in 1922. In elections the following year, thirty-four parties or groups presented candidates. Pašić and the Radicals gained considerably, and the Croatian Republican Peasant Party rose from fifty to seventy deputies, all of whom refused to take their seats. Had they done so and combined with Pašić's opposition, he could have been defeated.

Indicative of continuing political uncertainty was the fact that in 1924 Yugoslavia had five cabinets, four headed by Pašić and one by Democrat Ljubomir Davidović. The latter, like Pašić and others, negotiated with the Croats in an effort to achieve their political cooperation. These attempts resulted in splits in several parties, and made more difficult King Alexander's attempt to promote an all-party cabinet. (Although Alexander had exercised the royal powers since 1914, he was made king only when his ailing father, Peter I, died in 1921.)

While Pašić was seeking to govern with unstable coalitions, Radić worked furiously at home and abroad to realize Croatian independence. He sent letters and memorandums to international conferences, to the League of Nations, to Britain's Lloyd George, and to the U.S. Congress. On several occasions he also indicated his willingness to seek an honorable agreement with the Serbian people, but not with Pašić and the Radicals. On the anniversary of the fall of the Bastille in 1923, he said he viewed Croatia as imprisoned in a "Serbian Bastille," and he referred to the queen as "Madame Pompadour."

Learning that the government intended to arrest him, Radić and some of his collaborators fled in June 1923 with false passports obtained in Hungary. He went to London, Vienna, and Moscow, and seemingly got no encouragement except perhaps in Moscow, where he aligned his party with the Peasant International, at that time controlled by the Third (Communist) International. This act, which aligned his

Catholic party with atheist Communists, convinced the Yugoslav government that Radić had desperately brought into question the unity of the state. In fact, it eased the position of the Radicals, who had been viewed by some as too harsh on the Croats.

Radić returned to Zagreb in August 1924, believing that it was safe to do so with the Davidović cabinet in power. When that cabinet fell in the closing weeks of the year, however, the new Pašić-Pribićević government decided on new elections and harsh measures against the antistate Croatian Republican Peasant Party and its leaders. The legal basis for the latter was the law on the security of the state, and the main evidence was Radić's joining the Peasant International. By these actions the cabinet aimed to secure an electoral victory and, if possible, a reduction in the number of Radić deputies.

Radić and several of his party comrades were arrested in January 1925. New elections were held in February, with a large turnout of voters. The Radicals made modest gains (from 108 to 123 seats); with their allies they now had 160 seats out of 312. Radić's party actually gained in popular votes, but lost three seats. Still, its 67 seats constituted the largest group in the opposition's total of 152.

These elections and the cabinet's action against Radić had some unexpected consequences. The success of the Pašić-Pribićević government in winning a majority was a Pyrrhic victory. Several of the Croatian deputies were in jail, although even before the elections charges against most of the Croatian leaders had been dismissed. The investigation of Radić continued, but King Alexander was reluctant to see him brought to trial for fear that such an act would be viewed as political persecution. Slowly the politics of concert and accommodation was replacing the politics of rigidity and hostility.

Alexander paved the way for an agreement by permitting negotiations with Radić while he was in prison. The result was that Radić was amnestied, and a Pašić-Radić cabinet came into being. Some viewed this as capitulation by Radić. In fact, it was a victory, since he and his party leaders went directly from prison to posts in the government. It was also a victory for the palace. For the Radicals it meant a loss of prestige, because they were forced to retreat from a policy of firmness to one of sharing power with those they had imprisoned.

The first public sign that the Croats were prepared to depart from their past position was a statement in parliament in March 1925 by Radić's nephew. In it he said: "We recognize the total political situation as it is today under the Vidovdan Constitution with the Karadjordjević dynasty at the head." Also significant was the dropping of the word "Republican" from the party's name. The party also renounced its connections with the Peasant International.

Radić had apparently been ready to reach an agreement the previous year, when he told some of his supporters that he was willing to recognize the monarchy, but he had added: "For God's sake, give me time to turn my automobile around. . . . I cannot change all of my politics at once." Subsequently, in a private letter from jail, he had written: "We did not become monarchists out of fear, or in order to get out of prison, but because, as a party, we entered into the constructive period of our work."

Negotiations leading to the agreement were protracted,

but it was formally signed in mid-July 1925. The press referred to the new cabinet as R-R (Radical-Radićist). In it the Croatian Peasant Party headed four ministries, though Radić did not take his place until November. Among the first acts of the new cabinet was the dropping of criminal charges against Radić and other Croatian Peasant Party leaders.

After his release from prison, Radić had nothing but praise for Alexander. To reporters, after an audience with the king, Radić said: "Our king is our man. . . . There are no people like ours, and . . . no king like ours." When Alexander visited Zagreb in August 1925, Radić composed a hymn to him.

His change of policy was no doubt motivated by a lack of success abroad, as well as by the fear that Croatia might be cut loose ("amputated" was the word often heard in Serbian circles), leaving it prey to the ambitions of Italy's Mussolini.

A Serbian opposition leader and dedicated Yugoslav who cooperated with the Croats, Dragoljub Jovanović, attributes a great deal of the earlier difficulty to Radić's volatile personality. He thought Radić a fascinating politician: no one stung Serbia more, yet no one glorified it more; no one battered the state, the monarchy, and the dynasty more, yet no one lauded them with more enthusiasm. According to Jovanović, Radić changed his thinking toward everyone and everything, parties and individuals: he could be quiet and withdrawn or angrily outspoken; he could hate or like without reason.

Pašić, for his part, had known an agreement with the Croats was unavoidable sooner or later. His belief was strongly influenced by Alexander's decision to follow a pol-

icy of appeasement and conciliation. Pašić was no less per-
suaded by the results of his policy of firmness, which instead
of crippling Radić's movement actually solidified its control
over the Croatian masses.

Reactions to the agreement varied. Most Radicals accepted it. Others doubted Radić's sincerity, and were concerned when the agreement resulted in a break with the
Pribićević Independent Democrats. They had reason to be
anxious about the latter, because Pribićević was personally
hurt and soon opened a bitter campaign against the agreement. Most of the other political parties were divided. The
Slovene People's Party took a moderate position that enabled it subsequently to enter a cabinet with the Radicals.
Inside the Croatian Peasant Party, division prevailed; some
even accused Radić of treason. For his part, Radić defended
the agreement, merely suggesting at times that modifications
in the political system might be needed.

The promising future after the Pašić-Radić agreement soon
gave way to doubts as the cabinet entered stormy seas toward
the end of 1925. Disillusionment came from Radić's public
statements, from discord in Radical ranks, from charges of
corruption brought by Davidović Democrats, and from the
king's conviction that a change in leadership was needed.

Radić criticized the cabinet's fiscal policies and attacked
individual ministers, behavior generally impermissible by a
member of the cabinet. Many Radicals who had applauded
Pašić's bringing Croats into the cabinet saw this as a move
by the old politician to stay in power at all costs. They were
anxious that he make room for new leaders.

Meanwhile, the cabinet's woes were aggravated by the
charges of corruption in high places, which included Pašić's

son. The allegations were made by Davidović Democrats, and subsequently by Radić. There was no proof that Pašić knew anything about the corruption or that he had tolerated it. On the other hand, there was evidence that some members of all parties had taken advantage of the chaotic conditions following the war to engage in questionable deals and shady practices. King Alexander apparently was encouraged to agree with Radicals who wanted to challenge Pašić's leadership. When Radić resigned in April 1926 to protest Pašić's determination to postpone a parliamentary inquiry into corruption, the cabinet fell. It was the end of Pašić's long political career.

A new Radical-Radić cabinet proved unstable because of disarray in Radical ranks and Radić's penchant for delivering speeches critical of the government and his cabinet colleagues. A measure of the continuing political instability can be found in the fact that four separate cabinets were formed in one year. In the midst of the uncertainty the old political warrior Pašić was invited to the palace, because the king felt the need to explain the existing political course to him. All accounts agree that the session was a stormy one. That evening, while eating dinner, Pašić suffered a massive stroke, and early the next morning he died. This man, who first served as prime minister in Serbia in 1891, is now seen as one of the most astute figures in European history.

The cabinet's problems with Radić were of a different order from those within the Radical Party. Radić frequently criticized government policies as if he were a member of the opposition. He and other Peasant Party leaders, in the press and in speeches, constantly referred to the Croatian nation and its individuality, and to Croatian rights. The result was that all cabinets in which Croatians participated were in continual crisis.

New elections, in which twenty-seven parties presented candidates, were held in September 1927. Results were mixed, with most parties losing strength. The most significant outcome was the formation of an unlikely alliance between two bitter enemies: Stjepan Radić, a Croat, and Svetozar Pribićević, a Serb who had earlier collaborated with the Radicals. In November 1927, they concluded an agreement that established the Peasant – Independent Democratic coalition. They were never able, however, to establish a common front with other opposition groups to form a cabinet. Instead they engaged in uncompromising attacks upon those in power, and in parliament their obstructionist tactics led to paralysis. The king offered Radić an opportunity to form a cabinet. He accepted, but the refusal of the Serb-led Radicals to enter his cabinet led to failure.

Radić and Pribićević launched new attacks on the governing coalition cabinet. In separate audiences with the king, they proposed a "nonparliamentary solution," that is, a cabinet headed by a general. Pribićević asserted openly that a "dictatorship is better and more honorable than this false parliamentarism."

It is clear to historians today that the Radić-Pribićević coalition led step by step to increased polarization in the political system. Irreconcilable and unbridled struggles between this coalition and the cabinet were the order of the day. Crude remarks by Radić and his friends were often answered in kind by Serbs. In parliament, Radić called ministers thieves, bandits, outlaws, and murderers. The cabinet seemed powerless and irresolute, although it sometimes excluded him from sessions for a day or more. In reflecting on these events many years later, Radić's successor as leader of the Croatian Peasant Party, Vladko Maček, observed that the Radić-Pribićević coalition practiced "systematic obstructionism in the National Assembly, aided by the quite liberal order of procedure."

By early June 1928, the atmosphere in sessions of the parliament became almost unbearable. Sharper, more bitter conflicts between Radicals and those of the coalition occurred more frequently. The climax of the crescendo of attacks and insults came on June 20, when a deputy from Montenegro named Puniša Račić drew his pistol in parliament and shot four Croatian deputies, including Radić. Two died almost instantly and a third soon after in the hospital. Radić was wounded but recovered and resumed his activities. Yet he died two months later, from what experts agree was a secondary infection associated with his diabetes.

The assassinations meant more than a breakdown of an uneasy consensus: they signified the end of the hard-won political system. The Montenegrin was a Serb, of course, and Serbs were blamed for his self-impelled violence. Members of the Radić-Pribićević coalition left Belgrade, vowing not to return unless the system was changed. Alexander engaged in last-minute efforts in the latter part of 1928 to preserve Yugoslav democracy as a viable political system, but he had no chance to succeed. There was again talk of "amputation," of cutting Croatia out of Yugoslavia. The king appointed as prime minister the head of the Slovene People's Party, Anton Korošec, hoping that he might be able to heal the breach between the Croats and the Serbs. This move proved fruitless. He then talked with Pribićević and with

Radić's close associate Maček, but they saw no hope unless the constitution was amended to change the system. The governing coalition did not consider this a responsible solution.

Alexander consulted other political leaders and the leading constitutional expert, Professor Slobodan Jovanović, and on January 6, 1929, announced that as a result of his far-flung consultations he had concluded that there was no parliamentary solution that would guarantee the preservation of full state and national unity. Consequently, he was assuming personal power. Thus efforts to find a viable political system for the Kingdom of the Serbs, Croats, and Slovenes through parliamentary democracy had failed.

Two issues in the Serb-Croat struggle can now be seen clearly, in the light of subsequent history. Did the Croats have a chance to live within a safe—civilly and internationally safe—independent state? That is highly doubtful. Yugoslavia was a large country of sixteen million people. Croatia would have been a small country of fewer than three and a half million people, bordered by Italy, Austria, and Hungary, all former masters. Would present-day Croatia have been able to claim the large territory it proclaimed as its own when declaring its independence in 1991? Certainly not. The irony is that in the First Yugoslavia, Croatian territory expanded, and that Yugoslavia under Tito further increased it. Great Croatia is as much an issue in strict historical terms as is Great Serbia.

Any attempt to find the reasons for the unbridgeable discord and for the ultimate failure of the system is fraught with difficulties. Perhaps the simplest answer to an exceedingly complex question is that the Serbs, Croats, and Slovenes had never in the past lived in a single common state. When they attempted it, their problems of interrelationship were overwhelming and actually beyond their capability to settle within democratic institutions that were for many new and unfamiliar. Except for Serbia, parliamentary government was a new practice in the political life of the South Slav people; it had never been practiced in Austria-Hungary or in areas once part of the Ottoman Empire. Nevertheless, it *is* possible to identify certain specific factors that contributed to the failure.

The creation of Yugoslavia represented a huge political and emotional step, and it was based to a considerable degree on illusory hopes and unrealizable idealism. The most perplexing problem was the latent distrust of Serbs among Croats, which had been systematically cultivated and spread by the enemies of a common state. Efforts on the part of Serbs to allay fears of Serb dominance failed to pacify the Croats, who carried a burden from the past. Having lived as a minority in Austria-Hungary, they were understandably defensive, and it would seem not unusual that they would engage in using the same means of passive resistance that they had employed before. Their psychological inheritance as Slavic subjects in an Austrian and Hungarian empire made it difficult for them, once freed of that empire, to accept the principle of majority rule within a state with an elected parliament that chose a cabinet.

The Serbs were seemingly ill-equipped to deal with the feeling on the part of many Croats that they had been tricked or misled into making a bad deal. Although the Serbs were aware of the historical, religious, cultural, and political differences between themselves and the Croats, they did not realistically assess and appreciate the difficulties ahead. They perceived the common state as an extension of their own history—not so much an aggrandizement as a continuation of an already existing unity: expansion, yes, but not dominance. Moreover, some Serbs believed that understanding would result from democratic discussions in a society where complaints and criticisms could be freely aired. When the air turned foul, they, too, were resentful: Living in a common state was supposed to resolve problems, and time would be a great healer. It was a naive hope.

Another source of difficulties was the Serbian army. When independence came, accompanied by Serbian troops, the Croats remained more loyal to their priests and their city leaders than to these peasants who were liberating them. Moreover, the Serbs evoked a certain resentment because they believed that their sacrifices—in seven wars in the past century and a quarter—should not be taken lightly. Their officers did not mention it, but foreigners commented widely on the fact that Croats had fought in the Austro-Hungarian army as enemies of the Serbians and Montenegrins. French Premier Georges Clemenceau, for example, stated that he would never forget that the Croats had fought on the side of the enemy.

In addition, the multiparty system made it impossible for one party to obtain a majority in parliament. Hence every cabinet was forced to depend on coalition support, and was never strong enough to undertake far-reaching or imaginative programs.

The nature of the party structure was such that it was unclear who was speaking for whom. The one exception was the Croatian Peasant Party, a major regional party. Serbian parties sought to organize nationally—in part because Serbs were settled in almost all parts of the country—but they had little success outside Serbia. At the same time, they were badly divided and never spoke with one voice as a Serbian party. Serbs in Montenegro and Macedonia and Bosnia-Herzegovina had their own agendas. Had all political parties been regional only, perhaps, with hard bargaining, they could have reached some consensus. Two small regional parties, the Slovene People's Party (Slovenia was little burdened with minority peoples) and the Muslim Organization centered in Bosnia-Herzegovina, were, indeed, able, in the confused and uncertain circumstances, to extract from coalition cabinets many benefits for their respective regions.

The Yugoslav Communist Party's stand on the nationality question changed. Initially, it saw the Croat-Serb conflict not as a religious or nationalist struggle, but as an economic-political one, a conflict between the Croatian and Serbian bourgeoisie. Later, it viewed Yugoslavia as a Versailles creation that should be destroyed.

Another factor was the role personality played. When it came to dealing with the Croats, the Serbs listened to Pribićević, who had lived and worked with Croats within the Austro-Hungarian Empire. He told the Serbs that the Croats would respect a firm hand, but their acting on his advice produced intense hostility toward the Serbs. Pribićević often changed political direction. He began as a worshiper of monarchy and unitarism and later became a republican and a federalist. As an ally of the Croats in the last years of parliamentary democracy, he played a critical role in frustrating the political process.

Similar things can be said about Radić. He did not permit his delegates to take part in constitution-making, and he boycotted parliament after the constitution was written. When Pašić succeeded in getting him to recognize the monarchy and the political system and join the cabinet, it did not take long for him to become an obstructionist. He changed his thinking about everyone and everything, which offended his natural allies, particularly the anti-Radical Serbs. Widely read, he was the one real intellectual in his party, yet among peasants he tried to portray himself as a man of the soil by attacking intellectuals as an unproductive class. When he was among them, he made the sign of the cross, but he did not hide his dislike of the Catholic clergy. He often gave the impression of being a great actor. He was a born demagogue, though one who sincerely believed in Croatian independence.

The Serb Dragoljub Jovanović, the leftist Agrarian leader who spent most of his political life opposing the Serbian Radicals and collaborating with the Croats, in his later years made some caustic comments about Pribićević and Radić. They "competed to see who could offend Belgrade and the Serbs more," and "who would minimize their [Serbs'] credit for liberation and unification more . . . so that some well-intentioned men momentarily approved that unforgivable crime" in the Belgrade parliament, "as well as the king's taking power into his hands." Of Pribićević, he observed: "He brought injustice to that land [Serbia] that really did the most for that large state that had been his ideal."

The Serbian Radicals, who headed most of the cabinets in the pre-1929 period, had their faults, too. They did not know how to deal with the Croats or were convinced that there was no compromise that would satisfy their demands and save Yugoslavia as an integral state. They were never in a majority position, in large part because of the divisiveness among the Serbian parties, and, confused and resentful, they were powerless to seek programs that might have had more promising results.

Internal Boundaries Established
by King Alexander in 1929

4

King Alexander's Attempts to Save Yugoslavia

Alexander Karadjordjević was baptized in the fire of politics very early. At age twenty-five, in 1914, he began to exercise the royal powers as prince regent of Serbia, shortly before his nation was attacked by Austria-Hungary. He was the nation's head in the trying days of World War I, when he and his aging father accompanied the Serbian cabinet of Nikola Pašić and remnants of the Serbian army in retreat across the Albanian mountains in the winter of 1915–16. After a year on the Greek island of Corfu, they returned to fight on the Salonika front in 1916, and were in Belgrade in 1918 to meet with the representatives of the National Council from Zagreb to proclaim the coming into being of the Kingdom of the Serbs, Croats, and Slovenes.

Alexander was a Serb, but also a sincere Yugoslav. His belief in a unified country could be seen even in the names he gave his sons. The first was given the name of his Serbian grandfather, Peter I; his second son was named Tomislav, after Croatia's first king, in the tenth century; his third received the Slovenian name Andrej.

Young and impatient, Alexander watched the provisional parliament as it stumbled along for two years, involved in every divisive issue before it could agree on an electoral law that would bring the constituent assembly into being. In the midst of enormous domestic and foreign uncertainty, the democratic process was characterized by sluggishness and instability. Some said that Alexander's military training in Russia inclined him to impatience with parliamentary institutions and their dissensions, delays, and dubious compromises. But as he listened to the interminable squabbles, criticisms and counter-criticisms, charges and counter-charges, insults and counter-insults, he did not need military training or any other conditioning for his patience to wear thin.

He recognized the threat to the new nation of the Croats' demands, which he viewed as a desire on their part to create a state within a state, but he kept hoping that a solution could be found. Yet it seems clear that he did not grasp the seriousness of the alarm signal set off by the agreement in 1927 between longtime rivals Radić and Pribićević.

If he *had* understood the gravity of this threat, what could he have done about it? As a constitutional monarch, what could he have done about parliamentary obstructionism? He did not cause, but also could not have prevented, the shooting in parliament; yet what could he have done about that? After the assassinations, he waited six months to see if the parliamentary system could right itself, all the while hearing important politicians whisper in his ear that a military man was needed. With more than fifty years of historical perspective, many observers now conclude that his decision to assume authoritarian powers in January 1929 was inescapable.

When he took personal power, Alexander explained that

he was motivated by the need to preserve national unity and the integrity of the state. To this end, he abolished the Vidovdan Constitution and dismissed parliament. Political parties were abolished. The standing law on the "protection of the state" was strengthened by actions against divisive organizations, chiefly those based on religion or nationality. He implied that the new dictatorial regime would be of brief duration, and set forth specific tasks to be accomplished "in the shortest possible time" by new laws to make the king the wielder of all political power, to institute censorship, to abolish local self-government, to ensure the security of the state, and to create a special court for political crimes.

Alexander chose to exercise his rule through a cabinet headed by General Peter Živković, who had earlier been commandant of the Royal Guard. Aside from Slovene Anton Korošec, no other major political-party leader was in the new cabinet, although some distinguished members of smaller parties did have cabinet posts. Most of the ministers had past parliamentary experience. The most important economic posts were given to Croats. Živković noted that the cabinet's task was limited in time, because as soon as the basic program was carried out "the royal government will proceed to study and implement measures for entrance into a healthy democratic . . . and full constitutional life."

The coming of the dictatorship did not evoke surprise at home or abroad. Party leaders believed that Alexander's personal rule would be temporary, as the king had indicated. Croatian Peasant Party leader Vladko Maček greeted the king's proclamation with: "The vest has been unbuttoned!" This was apparently a reference to the words of an old Magyar statesman, who purportedly said: "If the vest is buttoned the wrong way, the only thing to do is unbutton it and button it again the right way." Most believed that Maček was suggesting, as he had to Alexander earlier, that a new constitution, and hence a reorganization of the state, was needed.

Whereas the Croats welcomed the abolition of the Vidovdan Constitution, the Serbs were not happy, partly because their nineteenth-century struggle for parliamentary democracy had taught them that such a move would be a step backward. Moreover, they did not like having a military man as prime minister. Political party leaders in Serbia, Montenegro, and elsewhere had nothing favorable to say about Alexander's taking personal rule.

Maček and his associates soon developed doubts about the Živković cabinet, but they were reluctant to take any action, being convinced that the Serbs would actively oppose it. They were right. As early as April 1929, Democrat Ljubomir Davidović tried to talk with Maček, but the latter wanted nothing to do with Serbian politicians. The first Serbian politician to oppose the regime publicly was Dragoljub Jovanović, the leader of the leftist Agrarians. Arrested several times, he was first found not guilty, but subsequently was confined to a small village in the interior. Other leaders went underground and resorted to distribution of secret leaflets and mimeographed messages. Ironically, limitations on the press were felt more keenly in Serbia than in Croatia, but the Agrarian Party did publish an illegal newspaper, which appeared under different names.

During midsummer of 1929, after Maček and his comrades realized that the Živković regime would be more than temporary, they endeavored to take the Croatian cause abroad. Their efforts, as with Radić's some ten years earlier, met with singular lack of successs. Their moves obviously did have a negative impact on Serbian opposition leaders, who began to have serious doubts about Maček and his followers. Moreover, their moves disturbed their coalition friends, the Independent Democrats, who lost nearly all of their eminent Slovene members.

Unlike Maček, his partner Pribićević wanted to establish connections with the Serbian opposition. He traveled to Belgrade in May 1929, despite warnings of danger from Zagreb's chief of police. On his arrival at the railway station in Belgrade, he was promptly arrested and sent to a small town in Serbia. This move created confusion and fear, but little was heard publicly, except from Davidović. Pribićević was never formally charged or brought to trial. Following a hunger strike and certification by a panel of doctors that he needed to travel to Czechoslovakia for treatment, he was given a passport in July 1931. He never returned, but continued his fight against Yugoslav regimes from abroad until his death in 1936.

In December 1929, Maček was arrested for allegedly backing terrorist activities. He was found not guilty, although some of his colleagues were convicted. Not long after, a vice-president of the Croatian Peasant Party, Karlo Kovačić, joined the Živković cabinet, and a few months later four more prominent members of that party were included in a reconstructed cabinet. Another party vice-president, Josip Predavac, however, was sentenced to two and a half years in prison in connection with a bank failure in Zagreb.

Foreign reaction to Alexander's assumption of personal power was generally approving. His strongest support came from France and Czechoslovakia. Yugoslavia was a key partner in the Little Entente, a defense pact consisting of Czechoslovakia, Yugoslavia, and Romania that was unofficially sponsored by France. As for the United States, its attitude is shown by editor Hamilton Fish Armstrong, who wrote, after Alexander's death, that he "did not become dictator because he was avid for personal power," but because the "country seemed not far from civil war."

British official circles were reserved, but the influential *Times* wrote that it was difficult "to find fault with King Alexander because he dared not wait until confusion had become still worse compounded and until the fissure in Yugoslav unity had widened into gaping chasms." The British minister in Belgrade, Neville Henderson, noted in June 1930: "I find it difficult, in view of the ghastly mess into which the Parliamentary regime had brought the country, to see what other course the king had. . . . He had to take personal responsibility as the link of the Crown was the only bond between the warring factions." In his memoirs, published many years later, he was just as forthright. Alexander, he said, was democratically minded and regarded the dictatorship as temporarily in the best interests of Yugoslavia. "His concern was the unity, future welfare, and happiness of Yugoslavia as a whole, and no greater or truer patriot ever existed."

Alexander's position was a tortuous, and tortured, one. He had sworn to uphold the constitution, but he had also sworn to guard the integrity of the state, for which he had a strong personal feeling. For one thing, he had taken part in more actual fighting during World War I than had all the heads of other states, allies and enemies alike, put together.

There is little doubt that the country now faced a grave danger to national unity, possibly civil war. Even one of Alexander's bitterest critics, Pribićević, admitted that a large proportion of the population had little tolerance for parliament.

As was to be expected, the sharpest criticism of Alexander's move came from the Croatian separatist press and from government presses in Italy, Hungary, and Bulgaria, whose leaders had been hostile to the creation of a united South Slav state. Criticism from these sources tended, however, to bolster the king's support in friendly states.

Unfortunately for Alexander and his cabinet, the year 1929 brought the beginning of what was to become a worldwide economic depression, which had truly ruinous consequences for Yugoslavia. Many positive things were done to mitigate economic hardship, such as the moratorium on peasant debts, but these seemed not to be enough to compensate for authoritarian rule. In addition, certain economic measures, such as the closing of some teachers' colleges, embittered many intellectuals. The depression's effect on the regime was aggravated by divisions in the cabinet and erosion of foreign support.

By the fall of 1929, Alexander had done more than assume personal rule. He had changed the name of the country from the Kingdom of the Serbs, Croats, and Slovenes to the Kingdom of Yugoslavia, to signify state and national unity and suggest symbolically the full equality of all the peoples of the country. He permitted the flying of only the Yugoslav flag, thus banning the Serbian and Montenegrin flags, among others. At the same time, he reduced the country's thirty-three administrative districts to nine, called *banovinas* (a

Croatian term, from the word "ban" or governor). They were named after principal waterways, all but one of which were rivers. The boundaries of the *banovinas* cut across traditional historical lines, thereby serving to deny the old regions or areas their separate national character. Some critics argued that centralism was thereby enhanced; for others, the change in the name of the state and the creation of the *banovinas* were the first steps away from centralism.

Despite all efforts, the regime became increasingly unpopular. Opposition leaders in Belgrade and Zagreb believed that agreement among the parties that "had their roots in the people" was necessary if the dictatorship was to be replaced, but they were unsuccessful in attempts to reach agreement on a common platform opposing the system. The Belgrade leaders wanted, first of all, an end to the dictatorship, a return to political freedoms and free elections, and the restoration of the parliamentary system. Maček and his colleagues wanted first of all agreement on the territory that would belong to Croatia and a guarantee of Croatia's constitutional rights. When a major part of the Belgrade opposition—the Democrats and Agrarians—showed a willingness to accept Maček's demand for a federal organization of the state, Maček refused to work with them unless the Radicals also accepted a federal solution. Thus, even in adversity, the opposition to Alexander's rule could not cooperate.

The outstanding leader of the Belgrade opposition was Ljubomir Davidović, a Democrat. He recognized that some form of federalism was necessary, but pointed out that implementation of such a solution faced many hurdles. For example, a simple three-way division (Serb, Croat, Slovene) was impossible because of the intermingling of Serbs and

Croats in certain regions, a consideration that might require seven or more units. For the Serbs, who made up about half the population of the state, said Davidović, this would not correspond to the simplicity and clarity of the state structures that for more than a century had enjoyed the support of the Serbian people in Serbia and in Montenegro.

The king had hoped to break down the resistance of the Croats, but instead caused them to unite in opposition. He hoped to see the Serbs more unified, but instead saw the disintegration of Serbian political life. After 1931, no Serbian political party was strong enough to represent a large segment of the electorate. In Croatia and Slovenia, the opposite was true.

Alexander's enemies sought to portray him as a power-mad ruler who welcomed a pretext to become dictator and had even arranged it. The weight of available historical evidence suggests, however, that he took the step without enthusiasm and with no plan to continue it beyond a brief period. Refusing to copy other European dictators, he did not assume the title "leader." Mussolini referred to Alexander's regime as "a porcelain dictatorship" and called it "a shame for all us dictators." Alexander developed no ideology, neither monarchical nor fascist nor socialist, and for the most part his regime was not bloody.

Once, he complained that Živković had refused his request for money from the military budget to be used to build a monument to the Unknown Soldier at Avala, just outside Belgrade. "And here I am a dictator and an autocrat, and my minister . . . answers, 'I will not give it to you.' "

In September 1931, Alexander formally brought his personal rule to an end and inaugurated a system that can best be described by a post–World War II term, "guided democracy." He wanted to return power to the people, but with some limitations, to avoid the evils of the pre-1929 system—in other words, he wanted a qualified and constitutional system. Political freedoms would be returned gradually, in periodic doses. Ostensibly, this timing was necessary because the passions of particularism had not yet died down.

To this end, Alexander handed the Yugoslavs an octroyed constitution that proclaimed Yugoslavia a hereditary constitutional monarchy, with the king as champion of national unity and the integrity of the state. It provided for a representative system, but added an upper house, half of whose members could be appointed by the king. Legislative initiative was in the hands of the cabinet, which was responsible only to the monarch. Hence the constitution did not reestablish parliamentary government; indeed, the word "parliamentary" was omitted from the text.

The unitary system was retained, but with a degree of decentralization. The *banovinas* were referred to as self-governing units and had elected councils—but no secret ballot. The sanctioning of the unitary principle satisfied neither Croat nor Serb opposition leaders, but many prominent men in the old parties saw the new system as an improvement and supported it.

The constitution was soon followed by a new electoral law, and elections were held in November 1931. Voting was direct and popular, but also public and oral. In order to get on the ballot, a person had to be on a party list that had some support in at least half of the *banovinas*. The law also required a statement from candidates that they would not join religious, ethnic, or regional political-party organizations. In order to avoid the roadblock of the former multi-party system, the law provided that the electoral list that received the largest number of votes in a *banovina* would get two-thirds of the unit's seats in parliament. The other one-third would be apportioned among the remaining parties that polled a set minimum number of votes.

The formation of political parties was legalized, but, owing to procedural and substantive limitations, none were formed until after the first election. Applications to form parties required a specified number of signatures and proof of some strength in a large number of districts. No organization, political or not, was permitted if it was founded on a religious or regional base, or if it opposed national unity, the integrity of the state, or the existing social order.

Hence the cabinet's electoral list was largely uncontested. The Croat and Serb oppositionists could not agree on a common program, or a common statement about the new political system, so they decided to boycott the elections. The small Communist Party, operating underground, received instructions from its leadership abroad calling for armed revolt, but nothing happened. The most visible protests against the regime came from students of Belgrade University, but they petered out after leaders of opposition parties realized that success was not at hand and advised the students to cease their actions.

The first political party to emerge was initially called the Yugoslav Radical Peasant Union, subsequently changed to Yugoslav National Party, and, after the elections of 1935, to Yugoslav Radical Union. In 1934, the Yugoslav People's Party was recognized, and later a rightist party came into existence, known as the Yugoslav National Movement.

To many Yugoslavs, guided democracy was a continuation of the dictatorship, but in fact it provided for a more lively political life. Conferences and meetings called for the purpose of organizing political parties offered many opportunities to speak against the existing system. Yet it is clear that these gatherings, as well as parliament, provided only a limited outlet for dissent and expressions of discontent. On the whole, parliament in this period was a sterile institution.

Under guided democracy there were several cabinets, all but one headed by well-known former Radicals. Following the appointment of Democratic dissident Vojislav Marinković as prime minister, there were signs that liberalization was on the horizon. A feeling of greater freedom served to lessen opposition to the regime. Political leaders moved about more freely and issued statements to their followers. At public meetings there were discussions of political rights and freedoms, as well as criticisms of the constitution. But as hopes of getting the Croats to accept the new constitution waned, so did hopes for liberalization.

Alexander's efforts to establish contact with Maček and other Croatian leaders were unproductive. The Croats cautiously suggested that agreement might be possible concerning the monarchy, the integrity of the state, foreign policy, the army, and finance, but they were determined to avoid being concrete. For them, the Croatian question was uppermost; dictatorship and political rights, although important, were secondary. Maček did not really want to talk with the king, because he was determined not to participate in any governing coalition, even if he were invited.

At the same time, messages to Maček and his colleagues from Serbian opposition leaders seemingly fell on deaf ears, mainly because the Croats were convinced that the regime was in trouble and that by waiting they would achieve more than by compromise, a belief shared by most Serbian opposition leaders. Yet the latter were divided as to what concrete program to follow. Some were willing to accept federalism; others were not.

Both Croatian and Serbian opposition leaders, in their separate ways, became bolder in their activities. At first, the government sought to avoid confrontation and limited its actions to discouraging peasants from attending protest meetings and imposing small fines or handing out reprimands or warnings. When the Agrarians issued sharply worded pamphlets, suggesting action in the fight for civil rights and self-government, as well as the need to find a solution to the nationality problem, their leader, Dragoljub Jovanović, and several of his comrades were arrested and given prison terms of up to one year.

Maček began psychological war against the regime. In a letter to an English journalist, he asserted that it was necessary to go back to 1918, that is, to start all over again. He pointed out that the parliaments of the historic provinces, seven in number, would need to reach an agreement with the federal government concerning what powers would be vested in a central government. In April 1932, he told a French newsman that Yugoslavia was "like a sick man with an incurable disease who certainly will soon die. That death will liberate Croatia."

In November, the Peasant-Democratic coalition, which had continued to function (although not publicly) despite the prohibitions of January 1929, adopted a resolution allegedly intended for internal purposes. It soon found its way into the foreign press and then was published by the domestic press with caustic criticism. Among several demands was one that asserted it was necessary to go back to 1918 so that all peoples of the state could agree on a new organization of governmental affairs.

Government circles pounced on the request for a return to 1918, pointing out that this would serve only the enemies of Yugoslavia, especially the revisionists in Italy, Bulgaria, and Hungary. Except for the leftist Agrarians, Serbian opposition parties did not approve of this Zagreb resolution.

After Maček told an Austrian newsman that the Croatian question could not be solved within the boundaries of Yugoslavia, he was arrested for violating the security law, found guilty, and in April 1933 sentenced to three years' imprisonment. This action was greeted by public protests and, more seriously, by terrorist acts.

The Slovene People's Party issued a proclamation that seemed to echo the coalition resolution, but after the arrest of its leader, Korošec, the party issued a new declaration, which made it clear that its demands for a reorganization of the state should not be interpreted as an attack on the existing Yugoslav state and denying any connection with the Slovene federalists or Italian Fascists. The Yugoslav Muslim Organization, which had associated itself with the Peasant-Democratic coalition, had to pay only a police fine, because its resolution was milder than Maček's and because its leader was initially opposed to it. The Communist Party did not play any meaningful role.

Aside from a watered-down protest of Maček's imprisonment, Serbian and Croatian opponents of the regime could agree on very little. In short, Croatian opposition was *national*, whereas Serbian opposition was *political*.

After Maček's incarceration, relations between the partners in the Peasant-Democratic coalition deteriorated, partly because his colleagues, who were in charge of party affairs, veered toward a more extremist course. Moreover, the position of the Independent Democrats was made almost untenable by the antimonarchist and revolutionary rhetoric of their exiled leader, Pribićević. At the same time, Croatian Peasant Party leaders were trying to put as much distance as possible between themselves and Pribićević; he had burned all his bridges to the palace—a step the Croatians had no intention of taking.

In the meantime, two Radical prime ministers thought they could find a solution to the Croatian problem, but were unsuccessful. One put great hopes in legislation that would give the *banovinas* fairly broad self-governing powers, but cabinet discord prevented its being presented to parliament. Many cabinet members were convinced that they were seeing only a short-term, extraordinary situation, so they were not moved to action.

Alexander was keenly aware of the various difficulties and at times expressed personal frustration. He was not opposed to local self-government, but he was convinced that federalism would lead to constant jurisdictional conflicts between national and regional parliaments. Yet he was not pleased with the results of guided democracy. He also faced foreign-policy concerns, especially Italy's determination to make trouble in order to annex territory.

As can be seen, attempts to impose consensus from above,

through personal rule and guided democracy, were unsuccessful. The Serbs, because of their earlier successful struggle for parliamentary constitutional democracy, could not be satisfied by what to them was a step backward. In addition, the opposition to the cabinets of this period was ineffectual and unproductive, mostly because of Maček's intransigence, based on his belief that the longer the Croats waited, the better terms they would get. The result was political stalemate.

Before going to France in October of 1934, Alexander confided to several persons that after his return he would free Maćek and meet with him personally. He was convinced that the two of them could solve the Croatian question. That was not to be, because a plot by extremist Croatians to assassinate him succeeded shortly after he reached French soil. In Marseilles, on October 9, 1934, he was greeted by French Foreign Minister Louis Barthou and entered an automobile for an official parade. An assassin strode to the side of the car and shot both of them. They died almost immediately.

5

The Search for a Solution After Alexander

If the Croatian extremists who engineered Alexander's assassination expected Yugoslavia to disintegrate, or the existing regime to collapse, they were grossly mistaken. The king's murder shocked the nation. Alexander's alleged last words, "Safeguard Yugoslavia," struck a positive response and tended to endow his memory with a heroic political mystique. Although most observers doubted that the king had uttered any words on being shot, no one thought that these words misrepresented his sentiments.

Because his eldest son, Peter, was only eleven, royal powers were assumed by a regency of three men, as provided by the king's will. Prince Paul Karadjordjević, Alexander's first cousin, in effect soon became the regency. He confided to a close associate of the king's that he had not been nurtured for the role, that politics had never interested him, that Alexander had never instructed him in anything. Thus he found himself in a delicate position, in terms of both domestic and foreign affairs. Nevertheless, he soon demonstrated a determination to bring about change. He realized that he must work through political leaders, and he and they were in general agreement that solving the Croatian question was crucial.

Paul wanted to release Maček from prison but did not want this action to reflect adversely on the late king or his policies. After the existing cabinet resigned, in less than three months, Paul selected a dissident Radical, Bogoljub Jeftić, to be prime minister. One of the first acts of the new cabinet was to recommend amnesty for Vladko Maček, who was promptly released.

Jeftić declared himself a loyal defender of Alexander's program to safeguard Yugoslavia, and maintained that the 1931 constitution provided the necessary framework for the future. He attempted to include representatives from several established political parties in the cabinet, but was not too successful. He did appoint an eminent man, Milan Stojadinović, a financier and member of the Radical Party's central committee, to be minister of finance, and he also named three Croats.

The initial acts of the Jeftić cabinet appeared promising. In addition to amnesty for Maček, it succeeded in lowering interest rates at all banks, appropriating money for a public works program, and declaring a brief moratorium on peasant debts. Jeftič believed these measures would help him in the upcoming election. The electoral law was modified so that more parties found it easier to present lists of candidates, but the open ballot remained.

The Jeftić government was surprised when the Agrarians reached an agreement with the Peasant-Democratic coalition to contest the government in the elections. Insult was added to injury with the selection of Maček to head what became the United Opposition list. These actions by the Agrarians and the Democrats—two Serbian-based parties—caught the Serbian Radicals off guard, as it did the Slovene People's Party and the Yugoslav Muslim Organization, although the last soon joined the United Opposition list. The party created as the instrument of guided democracy, the Yugoslav National Party, was, in 1945, a group of officers without an army.

In the event, the Jeftić list received a little over sixty percent of the votes; the United Opposition—without ever issuing a common program—over thirty-seven percent. The latter contested the official figures, but thirty-seven percent was a notable success, particularly since the voting was not secret. Under the complicated electoral law, the Jeftić list received 303 seats in the parliament; the United Opposition, 67.

The Jeftić victory was more apparent than real. Although the contesting of the election represented a partial step toward political consensus—since it meant that the election was recognized as a reality—the action of the Peasant-Democratic deputies in not taking their seats repeated the negativism of earlier years. The victory was also diminished by the fact that a large mass of voters had openly cast their votes against the cabinet's list.

Disagreements in the cabinet soon led to Jeftić's resignation, in June 1935. It was said that he was a man of good will and sincere intentions, but without vision or a political plan. Some said that he behaved as if the Croatian question did not exist, because for him there were no Serbs, Croats, or Slovenes—only Yugoslavs.

There is reason to believe that Paul and his advisers came to the conclusion that the elections had been divisive, and consequently it was important to have a new cabinet composed of groups that had not been involved and presumably were not in conflict with the two opposing camps. This meant constructing a cabinet from Radicals, the Slovene People's Party, and the Yugoslav Muslim Organization.

Paul's new choice for prime minister was Stojadinović, a Radical who had joined the previous cabinet but without his party's consent. Before appointing him, Paul consulted Maček, who indicated that he wanted to be cooperative, but at the same time demanded a neutral cabinet that would immediately hold elections for a constituent assembly. The regent and many constitutional experts believed that the constitution could not be changed while King Peter II was still not of age. The Stojadinović cabinet was joined by men from the largest Serbian, Slovene, and Muslim groups, and by three Croats, signifying some support from Croatia. The cabinet leaders decided to unite in one political organization, the Yugoslav Radical Union, in order to create a majority in parliament.

Stojadinović, as a Radical stalwart and an opponent of personal rule, led people to believe that the authoritarian regime was near its end. Some promises of liberalization were postponed, but generally speaking there was a good deal of free speech, and the press was free enough to give full coverage to speeches and pronouncements by opposition politicians. Maček was able to give a fighting political speech in the center of Belgrade.

On the nationalities front, Stojadinović asserted that he wanted to build an atmosphere of confidence among Serbs, Croats, and Slovenes. In that atmosphere, he said, it would be easier to solve the Croatian question. Despite the expression of such noble sentiments, and although he was to be in power for three and a half years, he did not deal finally with the problem. The ranks of Croat separatists grew, and even in Maček's own party anti-Yugoslav voices were increasingly heard. Although Maček several times condemned separatist activities, many in his party thought him a procrastinator and a miserable leader. Moreover, he contributed to the growth of Croatian nationalism by permitting the establishment of Croat paramilitary organizations in both urban and rural areas.

During his tenure as prime minister, Stojadinović held one talk with Maček, in November 1936. Paul had told him he could talk about anything with Maček but must keep in mind that the constitution could not be changed and that federalism could not be accepted. Since Maček wanted to talk about little else, the meeting was unproductive.

If Stojadinović was in no hurry to solve the Croatian question, neither was Maček. With Europe in turmoil due to the rise of Fascism and continuing economic depression, the Croatian leader was convinced that international tensions would force a resolution of the Croatian question on more favorable terms than he otherwise might get. Moreover, weakened by the absence of his partner Pribićević, he was faced by defections from the latter's party, as well as continued insistence by Serbian oppositionists that the first order of business was the restoration of the parliamentary system. He was confident that negotiating with Prince Paul directly would yield greater success.

Complicating Stojadinović's position, although he did not seem to realize it, was the agreement in October 1937 between Serbian oppositionist forces and Maček's Peasant-

Democratic coalition. The agreement said that a new constitutional order was needed, that a new cabinet composed of all political parties should proclaim the end of the 1931 constitution and arrange for elections for a constituent assembly.

Stojadinović, feeling strong enough not to be bothered by this proclamation, was more concerned with the public business. He had eased the peasants' burden by restructuring and forgiving part of their debts; he had eliminated payment of taxes in advance; he had repealed certain trade stamp taxes and had lifted import duties on industrial machinery and automobiles. The country's balance of payments was now favorable, and government budgets showed surpluses.

Despite these positive developments in the state economy, the cabinet was shaken in 1937 by the reaction of the masses, particularly in Serbia, to the ratification of a new concordat with the Vatican. This agreement, which Stojadinović inherited, was largely the work of King Alexander. Several earlier concordats were in existence at the time of the formation of the new state in 1918, and a new one was needed. Since nothing had been done prior to 1929, Alexander initiated action by appointing the leading expert on concordats, Professor Charles Loiseau of France, to assist him. The draft agreement was ready at the time of the king's death, and the date for signing it had been set by Stojadinović's predecessor.

The patriarch of the Serbian Orthodox Church had given his blessing earlier, but when he and close associates read the fine print, they became convinced that it gave the Roman Catholic Church privileges that the other religious bodies did not have. Yet they did not react for more than a year. Why then did the concordat become an apple of discord when Stojadinović brought it to parliament early in 1937? The best answer seems to be that many of the prime minister's political opponents concluded that it could be exploited for partisan purposes.

Ironically, one of Alexander's reasons for negotiating the concordat had been to improve relations between his Catholic (Croat and Slovene) and his Orthodox subjects. Now it was being used for divisive purposes. When Stojadinović explained to an assembly of bishops why Alexander had negotiated it and that the regency wanted it ratified, he received a cold reception. To make matters worse, while discussion was taking place in parliament, Serbian Patriarch Varnava became ill and died on the day that parliament ratified the accord, amid rumors that he had been poisoned. In the light of widespread protests, in which the press did not play an objective role, Stojadinović decided not to submit it to the Senate. In his words, he "returned it to the desk drawer" where he had found it.

It is significant in this context to suggest that no people in Europe, with the exception of the Irish and the Poles, were more inclined to define themselves by Catholicism than the Croats. Catholics in Croatia were, and are, militant and quick to take arms. The Croatian frenzy in 1941 of killing Serbs and Jews and Gypsies was as much a religious act as a politically dictatorial or militarily terroristic act.

Stojadinović was in political trouble even before the turbulence over the concordat. Many citizens believed that he had dictatorial aspirations. At political rallies, some of his supporters wore green shirts, suggesting similarities to Mussolini's Blackshirts and Hitler's Brownshirts. He denied that he had such aspirations, but in recent years it has become known that he had told the Italian foreign minister that he wanted "to make his dictatorship popular." Foreign policy also became a bone of contention. In view of the lukewarm response in the West to the country's uncertainty after Alexander's assassination, the prime minister decided that Yugoslavia's foreign policy should be pragmatic and oriented to reach an accommodation with Mussolini and Hitler. This was highly unpopular in Serbia and among Serbs elsewhere.

Adding to Stojadinović's woes were signs of weakness in his own party. Although firmly in command of the Yugoslav domestic scene by late 1937, he had failed to conciliate the Croats or to rejuvenate the Radical Party as the representative of the majority of Serbs. Even though some members of his coalition remained, it was only out of calculated self-interest. He concluded that he could strengthen his forces through new elections.

These were held in December 1938. The prime minister's list did not do as well as he had expected. He polled fifty-four percent of the vote; the United Opposition got forty-five percent. Under the electoral law, his party got 306 seats; the United Opposition, only 67. Soon thereafter the old refrain was heard from Maček: the United Opposition's deputies would not take their seats in parliament. In view of the fact that the voting was not secret, and given the advantages enjoyed by the party in power, this was not a strong victory.

In any case, the victory was short-lived. Following a pseudo cabinet crisis in February 1939, Prince Paul replaced his prime minister with Dragiša Cvetković, a member of the outgoing cabinet. Why did Stojadinović fall? He himself rejected the argument that it was due to his failure to find a solution to the Croatian question. But there is no reasonable doubt that the refusal of the United Opposition deputies to take their seats in parliament, and the adoption of a resolution by the newly elected Croatians asking the great powers to intervene in Yugoslavia to assure Croats "liberty of choice and destiny," must have had a crucial impact on Paul.

Many years later, Stojadinović wrote that he believes the real reason was the desire of Paul and his wife, Olga, to become king and queen of Yugoslavia. He asserts that Paul—and constitutional experts he consulted—thought that the throne would become vacant if the heir died before he reached his eighteenth birthday. Stojadinović does not agree with this. He adds, however, that it was not accidental that at age fifteen Peter was permitted to drive a powerful Packard at high speeds on poor roads, and to pilot a plane when he was barely sixteen. Paul's biographers have written that there were several strong reasons for the dismissal of Stojadinović, including his foreign policy of accommodation with Germany and Italy, but that the alleged desire of Paul to become king was not one of them.

Be that as it may, the principal achievement of the new prime minister, Cvetković, was an agreement (a *sporazum*) with Maček, signed on August 20, 1939. A variety of complex factors led to this accord. Immediate events were ominous, among them the dismemberment of Czechoslovakia by Hitler after the Munich Pact and Hitler's exploitation of separatist tendencies in Slovakia (which later became a puppet regime under the Catholic priest Father Josef Tiso). Paul could visualize a similar danger to Yugoslavia. In addition, Maček, despite his binding agreement with the Serbian opposition not to deal unilaterally with Belgrade, believed that his best bet was to deal directly with Prince Paul, who had the ultimate authority and responsibility.

In January 1939, shortly before the change in prime ministers, Maček sought to exert pressure on Belgrade through a speech in which he said that the international situation was such that it was in the interest of the "Croatian and Serbian people that we find a solution within the boundaries of the state." Most Serbian circles were also convinced that international circumstances dictated compromise with the Croats. Maček, though he had sent a message to Paul saying he would not talk with any of the party leaders, finally agreed to talk with Cvetković, not as a leader of the Yugoslav Radical Union or as prime minister, but as a representative of the crown.

From Paul's point of view, Cvetković seemed a logical choice. Although not a figure with great standing among the Serbs, let alone nationally, he was known as having early been attracted to the idea of a Yugoslav state and had had contacts with the Croats already in 1928. Moreover, while a member of the Stojadinović cabinet, and with the permission of the prime minister, he had talked with Maček about a possible agreement.

To a special session of parliament in February 1939, Cvetković declared that finding a solution to the Croatian question was the cabinet's first order of business. He made it clear that he had the "greatest confidence of the royal regency," saying in effect that he was an agent of Prince Paul. In tackling the seemingly intractable Croatian problem, he faced twin difficulties: a Croatian leader willing to exploit international pressure on Yugoslavia to get the largest concessions for the Croats; and the determination of certain forces in Serbia, including Paul, not to let Maček exploit the situation to the detriment of Serbs and, especially, Serbia.

In talks with Paul, Maček and his emissaries had kept demanding a constituent assembly that would write a new constitution, and Paul had continued to oppose changing the constitution while King Peter was a minor. Moreover, the methods prescribed for amending the basic law required time-consuming procedures, which would produce prolonged debate, unwise in light of the tensions throughout Europe. At last, a legal and constitutional basis for allowing concessions to the Croats that would not formally alter the constitution was found in Article 116. This provided that in exceptional circumstances such as war, mobilization, disturbance, or rebellion, when the security of the state or the public interest was endangered, the king could "temporarily take by decree all extraordinary and necessary measures, independently of constitutional or legal provisos, in the whole kingdom or in one part." The only limiting provision was that such measures must subsequently be submitted to parliament for confirmation.

In parliament, initially, all groups had realized that resolution of the Croatian question must have top priority, and they therefore supported the cabinet. Some Serbian leaders were quick to point out, however, that if the cabinet was willing to abandon unitarism, then the most important question was not the Croatian question but the Serbian one. And United Opposition leaders, who had Maček's promise that he would not negotiate with Belgrade, felt betrayed.

Paul and Maček, facing these international and domestic circumstances, found themselves in agreement on the need to solve the Croatian question without delay. In advance of formal talks, Paul sought to determine through Maček's emissary the nature and extent of Croatian demands. Encouraged by Paul's readiness to reach an agreement, Maček, in four separate sessions, kept increasing Croatia's territorial demands. In April, Cvetković held several talks with Maček, made more and more concessions, and they thereby reached an agreement. It was not acceptable to Paul.

Several factors entered into Paul's rejection of the agreement. The Muslim leaders energetically opposed taking territory from Bosnia. Serbian leaders in Bosnia pointed out that a plebiscite would likely result in a million and a half *more* Serbs finding themselves inside a new Croatian unit. The military viewed any federal arrangement as weakening defense capability. Paul himself felt that a Europe poised on the brink of war changed the internal picture, and he hoped to use planned trips to Rome and Berlin to assess the situation further.

The fact that Paul retained Cvetković as prime minister after he had rejected the agreement indicated that he wanted to continue talks with Maček through Cvetković. The prime minister soon faced a serious challenge by followers of some Serbian members of his cabinet, and although he survived it, his support among Serbs was weakened.

Maček took the setback in talks with Belgrade in stride. The previous year, he had sounded out Italy and Germany, as well as France and Great Britain, to see what support he could expect if talks with Belgrade should fail. Following Paul's rejection of the April agreement, he sent an emissary to the Italian foreign minister, Count Ciano, to say he no longer intended to come to any agreement with Belgrade, and he asked for specific types of help to enable him to carry on separatist activities aimed at independence for Croatia. A memorandum of understanding was approved by Mussolini; among other things, it provided for an Italian grant of twenty million dinars. But when it was presented to Maček for signature, he refused it, saying he was once more involved in negotiating with Belgrade.

Subsequently, Maček claimed that the Italians had initiated the talks, but he admitted that the draft agreement did propose that in the event of war the Croatian Peasant Party was obligated to proclaim an independent Croatian state and seek the immediate backing of the Italian army. He denied, as claimed in Ciano's report, that the draft contained his pledge to mount a revolution in Croatia. Postwar scholarship seems to be in agreement that the version of the Italian foreign minister was more accurate.

Maček used the threat of foreign backing in the hope of forcing concessions from the cabinet and the crown. In an interview with the *New York Times* on August 1, 1939, he declared that if Croatia did not gain autonomy, it would secede from Yugoslavia, even though this would lead to civil war, and Croatia thereby might become a German protectorate. Paul and Cvetković and other political figures believed war in Europe was imminent, especially after Paul's trips to Rome, Berlin, Paris, and London. Their urgent aim now was to seek an agreement that might keep Yugoslavia out of the conflict. As a result, Maček was able to win from them major concessions. He and Cvetković resumed their talks in August, and on August 20 the sporazum was signed. Prince Paul quickly approved it, and on August 26 it was proclaimed. Maček signed it as both president of the Croa-

tian Peasant Party and head of the Peasant-Democratic coalition, even though this was contrary to his agreement with
his coalition partner, the Independent Democrats, and to his
obligations to the Serbian participants in the United Opposition campaigns of 1935 and 1938.

That same day, the cabinet resigned, and Cvetković
formed a new one, in which Maček became vice-premier and
five other members of his party got ministerial posts. At the
same time, it was announced that both houses of parliament
had been dissolved and that the crown had authorized the
cabinet to enact a new electoral law. Pending new elections,
the cabinet would rule by decree.

The sporazum created a large geographic unit known as
the Banovina of Croatia, which included the territory mainly
and traditionally inhabited by Croatians as well as most of
Slavonia (mixed Serb and Croat), Dalmatia (mainly Croatian, with a sizable Orthodox population), and parts of Bosnia-Herzegovina (Croats, Serbs, and Muslims). The definitive
boundaries were to be determined at the time of the reorganization of the state. Approximately one-fourth of the new
banovina's total population of four and a half million was
Serb. The Banovina of Croatia was to have its own popularly elected parliament, and a governor appointed by the
king. To a significant degree, therefore, the 1931 constitution was amended. Unitarism was not formally abolished,
but the autonomy of the Banovina of Croatia was tantamount to liquidation of it.

Considerable power, including fiscal, was to be dele-
gated to banovina authorities, though there would be some
overlapping. The central government would continue to have

control of foreign affairs, defense, communications, and transport. A constitutional court was to be created to decide cases of conflict in jurisdiction. Many things were left undecided, some were not done for several months. As an example, no electoral law was passed, so there were no elections, either nationally or for the Croatian parliament; and no constitutional court was established.

A royal decree, published as part of the agreement, declared that provisions concerning the Banovina of Croatia could be extended to other *banovinas,* including the amalgamation of territories or other alterations of boundaries. Presumably, this would guarantee that Serbian and Slovenian units could be established.

Croatian authorities soon placed restrictions on Jews who had fled from Germany, and they established internment camps for political opponents of the *banovina* regime. Moreover, the Cvetković-Maček cabinet took action against former prime minister Stojadinović, who, opposed to the sporazum, sought to organize a political party, using the name of the old Serbian Radical Party. After internment of less than a year, he was sent to Greece and turned over to the British, who took him to the island of Mauritius in the Indian Ocean.

It seems ironic that Maček and his party, who had succeeded in portraying themselves in the West as strugglers for democracy, accepted an order in which opponents of the sporazum were persecuted. It is also ironic that Maček, who in 1936 insisted that it was imperative to defend the position that Yugoslavia did not have a constitution, should play a role in its radical amendment in 1939. Noteworthy, too, is the fact that the open Yugoslav elections the Croats had condemned were in 1940 adopted in Croatia and praised as being democratic and honorable.

It is difficult to summarize the various reactions to the sporazum. Experts were sharply divided on its constitutionality. Since the cabinet continually postponed the election of a new national parliament, the agreement never received the ultimate constitutional sanction. Some viewed the agreement as creating a state within a state. Certainly, a large majority of Serbs believed that they would not be masters in their own house. Feeling that the Croats had obtained rights still denied to Serbs, they felt forced to demand the creation of a Serbian unit.

The reaction of the political parties was mixed. Those who had in the past collaborated with Maček felt betrayed, although they did not always show it immediately. Others were cautious, worried that the agreement would undermine the unity of the state. Some, including the Yugoslav Muslim Organization and the Slovene People's Party, were divided. Older Radicals were disinclined to demand a Serbian unit because this would signify acceptance of federalism. The Democrats, who had earlier accepted the principle of federalism, began pointedly emphasizing the Serb population of Bosnia-Herzegovina (almost one-half), Montenegro (almost total), Vojvodina (more than two-thirds), and Macedonia (at least two-thirds). The outlawed Communist Party, in some underground publications, first suggested that the Croatian problem had not been solved, then asserted that the agreement was unqualifiedly a bourgeois tactic to trick the working masses and the oppressed peoples of the whole nation.

The Croats, not denying the justice of the Serbian and

other demands for a reorganization of the state, felt that this should come only after the election of a new parliament that would ratify the sporazum. They feared that the establishment of Slovene and Serbian units, especially the latter, might preclude them from realizing their anticipated additional demands. This attitude created ill will outside Croatia, where the general view was that Croatia had already been given too much territory.

Parenthetically, it might be noted that Great Croatia keeps emerging in the history of our time. During World War II, the Croatian members of the Yugoslav government-in-exile sought to get a commitment on the boundaries of Croatia in postwar Yugoslavia; the Nazis vastly extended Croatian rule in 1941; and the Yugoslav Communists after the war made Croatia larger than it had been in 1939.

Not long after the 1939 agreement was signed, two — contradictory — documents depicted the Croatian attitude toward the sporazum. In a circular to party offices dated October 10, 1939, the Croatian Peasant Party leadership expressed complete satisfaction with the way in which Cvetković and the central government were implementing the agreement. It did criticize the uncooperative attitude of the bureaucracy inside the *banovina*. Viewing the agreement as the first phase in the total reconstruction of the state, it said that this was a calculated way of blunting separatist propaganda attacking Maček for betraying the idea of an independent Croatian state.

The other document, circulated at the same time, spoke completely contrary to the official one. It was labeled strictly confidential, but did not carry Maček's name or the party seal. Many were convinced that it had originated with the leadership, and some were sure that Maček had approved it. It asserted that in signing the sporazum the Croatian Peasant Party had not given up the idea of an independent Croatia. Rather, it was the first step toward its creation. Moreover, the agreement had achieved two goals: it had destroyed the integrity of the state — and consequently "Yugoslavia" should never be used, but simply "state union"; and the national government had been forced to move away from the idea of national unity, thereby destroying the foundations of Yugoslavia. Additional instructions told party members that in the central government they should always speak of "Croatian ministers," and the Banovina of Croatia should always be referred to simply as Croatia; Croats should always speak of a free and independent Croatia and of Croatian interests. Party members in the *banovina* were receiving similar directives "from Zagreb." From the autumn of 1939, many party organizations and local members followed the secret circular's instructions.

Some Croats maintained that the secret circular did not originate with party sources and therefore was not authentic. In 1990, however, a Yugoslav historian, Veselin Djuretić, who conducted research in the Soviet Union, reported: "On the basis of authoritative Soviet sources it is irrefutable that the prewar Croatian Peasant Party's divisive anti-Serbian and anti-Yugoslav circular is not a forgery . . . but an original."

Now came the horror that has so poisoned Serb-Croat relations since 1939. There arose among militant Croats a terrorist organization, the Ustashi, under the leadership of Ante Pavelić. Although Maček was probably inclined to seek a solution of the Croatian question within the Yugoslav framework, many of his followers in the provinces were under the increasing influence of the Ustashi, which was based in Italy. Ustashi attacks on Maček and the sporazum won for the organization more than a marginal following between late 1939 and early 1941.

As indicated above, as part of the sporazum package, a royal decree provided that the provisions concerning the Banovina of Croatia could be extended to the other *banovinas* by way of royal decree. To that end, commissions were actually set up to draft decrees creating Serbian and Slovenian units. The basic reason for failure to act was a dispute over the boundary between the Croatian and Serbian units. When war broke out, the Serbs, more concerned with national interests and national survival, generally favored putting the creation of their own *banovinas* aside.

The sporazum had been signed in August; World War II broke out in September. Poland was soon defeated, and Belgium, Holland, and France fell in the spring of 1940. At that point, for Yugoslavia available alternatives for alliance were unpromising. The rapid fall of France was a severe blow. The British were in no position to offer assistance, yet they expected the Yugoslav government and its army to rebuff Nazi pressure to adhere to the Tripartite Pact of Germany, Italy, and Japan. Croat and Slovene leaders were resolutely in favor of signing the pact; the Serbs, including Cvetković, were unswervingly opposed. Then in January 1941 Prime Minister Winston Churchill told Prince Paul that neutrality was not enough. Cvetković and Paul still believed that to enter the war meant to commit national suicide. Their one hope was to gain delays so that Hitler might leave them alone.

As it turned out, despite Hitler's impatience, the Yugo-slav leaders proved to be tough negotiators. Before signing anything with the Axis, they asked for concessions, which earlier Balkan signers had not. Hitler declared that what he was proposing to Yugoslavia was not in fact the Tripartite Pact. In a personal meeting with Paul, he took the same line, but also offered concrete guarantees.

The Yugoslavs signed a pact on March 25, 1941. At that time, Cvetković got three brief notes, signed by the German foreign minister. The first promised that "for all time" Ger-many would "respect the sovereignty of Yugoslavia." The second promised that the Axis powers would "not during the war demand of Yugoslavia the passage or transport of military forces through Yugoslav territory." The third stated that Italy and Germany would not ask Yugoslavia for any military assistance, leaving open the possibility that Yugo-slavia might at some point find it in its interest to offer help. Hitler agreed that the Yugoslavs could publish the first and third notes, but not the second.

The ink was hardly dry on the documents when, on the night of March 26–27, a military coup overthrew the Cvet-ković-Maček cabinet. The coup leaders declared young King Peter of age and ousted Prince Paul as regent. The new prime minister, General Dušan Simović, declared that the new cab-inet would abide by all international agreements that Yu-goslavia had signed. When noted Yugoslav scholar Slobodan Jovanović, a vice-premier in Simović's cabinet, examined the agreement signed with Germany, he asserted: "There is nothing here that could not be accepted." Yet the Cvet-ković-Maček cabinet was overthrown allegedly because of unwarranted concessions to Germany. Irony or not, the popularity of the coup was in part the result of a general lack of confidence in the existing cabinet.

Hitler ordered a massive attack; it ended the life of the First Yugoslavia. Consequently, there is no way of knowing if the sporazum was ever a possible first step in the establishment of a viable political system.

J. B. Hoptner, an American scholar who has studied the March 1941 coup, writes that the Allied leaders "failed to extend to Yugoslavia the patience and diplomatic restraint they showed to Sweden — despite the fact that Sweden, under conditions similar to those facing Yugoslavia, signed an agreement with Germany permitting a steady flow of German military traffic to pass over its borders." A subsequent study in Italy confirms the judgment reached by Hoptner: The coup was in the interests of Britain and, as it turned out, of the Soviet Union, but was suicidal for Yugoslavia, as Prince Paul and Cvetković had foreseen.

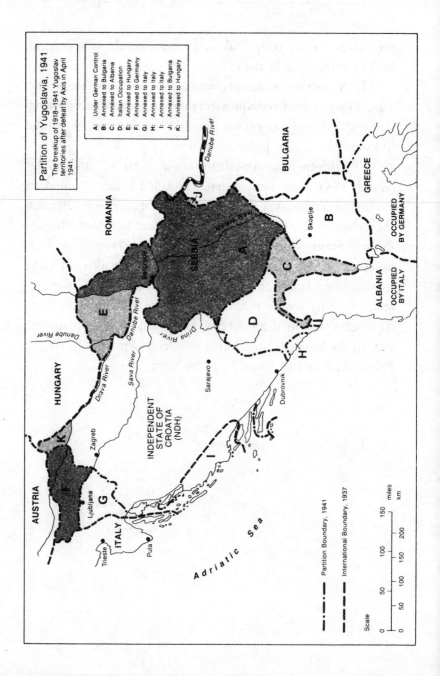

Partition of Yugoslavia, 1941
The breakup of 1918–1941 Yugoslav territories after defeat by Axis in April 1941.

A: Under German Control
B: Annexed to Bulgaria
C: Annexed to Albania
D: Italian Occupation
E: Annexed to Hungary
F: Annexed to Germany
G: Annexed to Italy
H: Annexed to Italy
I: Annexed to Italy
J: Annexed to Bulgaria
K: Annexed to Hungary

AUSTRIA
HUNGARY
ROMANIA
BULGARIA
GREECE
ITALY

Trieste
Pula
Ljubljana
Zagreb
Belgrade
Sarajevo
Dubrovnik
Skopje

SERBIA
INDEPENDENT STATE OF CROATIA (NDH)

OCCUPIED BY ITALY
OCCUPIED BY GERMANY
ALBANIA

Danube River
Drava River
Sava River
Drina River
Morava River
Danube River

Adriatic Sea

Scale
0   50   100   150   miles
0   50 100 150 200   km

— · · — · · Partition Boundary, 1941
— — — International Boundary, 1937

6

World War II and the Communist Rise to Power

The German-Italian attack on Yugoslavia at the beginning of April 1941 was short-lived. The Yugoslav army attempted to defend the nation on its northeastern borders where no mountains gave cover. It was soon driven back by tank divisions hardened in earlier campaigns. Belgrade, defenseless, was bombed. Resistance collapsed within two weeks, and the country was dismembered.

Less than a week after the outbreak of hostilities, while the Yugoslav army was still defending the nation, an Axis satellite, the so-called Independent State of Croatia, was established, under the leadership of Croat Ustashi leader Ante Pavelić, who entered the country with invading Italian troops. Slovenia was absorbed by Germany. Other parts of Yugoslavia were ceded to Italy, Hungary, and Bulgaria. The remainder of the state was divided into two occupation zones: one German, composed mainly of a rump Serbia under the civil administration of General Milan Nedić (who performed a role similar to that of General Pétain in France), and the other Italian, centered in Montenegro and areas northward, especially in Dalmatia.

The cabinet and young King Peter II fled, eventually

reaching London, where they remained except for an inter-
lude in Cairo, until the end of the war. Vice-Premier Maček
retreated to his home in Kupinec, Croatia, saying he wanted
to share the fate of his people.

The satellite Croatian state, ruled by the Ustashi, was
organized and functioned much like Mussolini's Fascist regime. Pavelić had to swallow a bitter pill, however: The
beautiful coastal area of Dalmatia, the pride of all Croats,
was annexed by Italy. He also was forced to accept as head
of state Italy's Duke of Spoleto, but this was pro forma only,
because the duke never appeared in Zagreb to assume his
throne.

Following Hitler's attack on the Soviet Union in June
1941, Pavelić declared war, and sent at least one military
division to fight alongside the Nazis on the Eastern Front.
He also sent thousands of workers from Croatia to Germany to assist in Hitler's war effort. After Pearl Harbor, he
declared war on the United States and Great Britain.

At home, Pavelić soon promulgated decrees to deal with
opposition, real or imagined. Under them, an enormous
number of his Serb subjects, as well as Jews and Gypsies,
were massacred or driven out. Some fled to Serbia and some
to the mountains south and east, where they subsequently
were part of guerrilla resistance movements.

Croat propagandists are still forced either to dissemble
or to quibble over the size of the massacre, and in the Western press in 1990–92 it has been almost impossible to find
an admission that a holocaust occurred in Croatia. Toward
the Serbs, Pavelić's regime used the formula "One-third we
will kill, one-third will be driven out of Croatia, and one-

third we will convert to Catholicism." The estimates of the number of Serbs killed vary from 300,000 to more than a million. A generally accepted figure is 500,000 to 700,000. The number of Jews killed was about 50,000, and the number of Gypsies was around 20,000.

As can be seen, most of the victims were Serbs. They were carted away to death camps, shot on the spot, or thrown wounded but alive into ravines. Others were herded into Serbian churches, which were then set on fire. Women and children were not spared.

Before long, ranking German officials in Croatia became horrified by the nature and extent of the killing, and came to believe that Pavelić wanted to kill all Serbs. They protested to his close associates, but these protests seem to have fallen on deaf ears.

In Serbia, not long after the Nazi occupation, some troops fled to the hills rather than surrender to the Germans. They began an uprising under the command of Colonel Draža Mihailović, and came to be known as the Yugoslav Army in the Homeland, or Homeland Army, but popularly as Chetniks (Cheta is a term used historically for irregular Serb fighters). After Mihailović's activities became known in the West, the Yugoslav government-in-exile promoted him to general and named him minister of war.

Initially, the Chetniks had no political aims other than to assist the Allies in defeating the Axis powers. The general viewed himself as a loyal servant of the Allied governments. But when the Nazis attacked the Soviet Union, and the Yugoslav Communists organized a competing guerrilla movement, he worked out a liberal democratic program and declared that the Yugoslav peoples should be allowed to determine their own political future once they were free to do so.

The Yugoslav Communists had held back until it was politically correct to fight the Axis occupiers. Then, under the leadership of Josip Broz, better known later as Tito, they sought to portray their National Liberation Movement, or, more popularly, Partisans, as broad and democratic. Knowing they had little following among the people, they concealed their real aims and hid the fact that they were Communists. Even the name of their leader was not generally known until many months after their guerrilla uprising was launched. They hoped that by disguising their ultimate political objectives, and with the help of the Soviets, they would be successful in gaining power.

The essential difference between the Chetniks and the Partisans was in their aims. The former pursued a national struggle for liberation from the occupier; the latter fought to seize power. Mihailović did not wait; Tito waited on the Soviet Union. Mihailović believed in freedom of choice for the Yugoslav peoples; Tito, in a dictatorship of the proletariat. These differences are the key to understanding all relations between the Communists and all other wartime groups within Yugoslavia, as well as relations between the Partisans and the Allies. It was unrealistic to hope for a united front.

Yet in the fall of 1941, Tito and Mihailović met to discuss a common front against the Nazis. The Communist leader's determination to preserve full autonomy of action in pursuit of his own agenda made agreement impossible, and the two men parted bitter enemies. The tragedy of the brutal struggles between them that followed was the more poignant because most of the recruits on both sides were Serbs. The Montenegrin Serbs suffered especially by being split between the two forces.

Mihailović succeeded in driving the Partisans out of Serbia when he realized what their real objectives were. They retreated mainly to the Montenegrin mountains in territory under Italian control. Some went into hiding in the fascist Croatian state, where they found abundant recruits among the Serbs who had fled from the Ustashi massacres. The Partisans never returned to Serbia in significant numbers until Soviet troops arrived at the Danube in late 1944.

In mid-1942, charges were heard from Tito's camp, via a radio purportedly broadcasting from the mountains in Yugoslavia but actually from the Soviet Union, that Mihailović was collaborating with the Germans and Italians. These charges were false, although some of Mihailović's supporters had fled to the Italian zone of occupation and may have asked for Italian help to save them from the Ustashi. Moreover, some Chetniks in the Italian zone were not under Mihailović's control, but had their own command structure.

At the time that Tito charged Mihailović with collaboration, he also sent his agents to talk with high-ranking German army commanders. They reached an understanding that the Germans would not attack the Partisans, so the latter could concentrate on fighting Mihailović. Hitler, however, vetoed this agreement. These talks have been confirmed by many sources, none more authoritative than Milovan Djilas, the Montenegrin who became one of Tito's four chief deputies before being imprisoned by him as a dissenter from Communist autocratic rule.

The military actions of the two groups against the occupier differed in a serious, even mortal, way. To counter Mihailović's guerrilla attacks, the Nazis began reprisals against the civilian population, taking fifty hostages for every German soldier killed and one hundred hostages for every officer killed. In October 1941, for example, they massacred between 4,000 and 7,000 men between the ages of sixteen and sixty in retaliation for the killing of twelve German soldiers. The action was repeated many times. Mihailović felt he could not justify bringing down upon the civilian population such merciless vengeance, and reduced his guerrilla activities. This was, moreover, in line with instructions from London to all underground movements to husband resources until the signal came that the West was ready to launch an invasion of Europe.

The Communist-led Partisans were not concerned about reprisals, believing they would force more people to flee to the hills, where they might be recruited into Partisan ranks. Admissions of this were to be found in the postwar Communist press. The Partisans also practiced brutal deception: In passing through territories hostile to them, they misled the Germans by leaving some of their booty behind, thereby inviting German punitive expeditions against innocent people uninvolved with the Partisans.

Among the Allies, the British took the lead in dealing with Yugoslavia's resistance forces. Liaison officers arrived by parachute to work, first, with Mihailović's forces and later with Tito's. They knew that both movements were pro-

Allies and that both were engaged in limited actions against the enemy, but they did not fully appreciate how much energy the Partisans were expending in trying to weaken the Chetniks. It is now known that the reports of the liaison officers, sent through British intelligence in Cairo, were being handled by a member of the British Communist Party, who passed on to London those favorable to Tito's Partisans and withheld those favorable to Mihailović. This is ably documented in two books recently published: David Martin, *The Web of Disinformation: Churchill's Yugoslav Blunder* (1990), and Michael Lees, *The Rape of Serbia: The British Role in Tito's Grab for Power 1943–1944* (1990). After a thorough study of British documents, Martin was able to show how much deception was practiced. Significant actions of the Chetniks against the enemy were often reported as the work of the Partisans. The editing of information by British officials was especially important at a critical juncture when the West was reconsidering its Yugoslav policy.

Prime Minister Churchill had, therefore, been receiving reports from his liaisons at Tito's headquarters that the Partisans were doing much more fighting against the Germans than were Mihailović's men. These reports had often been written before their authors had even visited the areas where Mihailović was operating. The result was that the Allies, or, more precisely, Churchill, changed policy toward the guerrilla movements in Yugoslavia in September 1943. British aid, which had been flowing generously to Tito, was now increased; that to Mihailović, which, ironically, had never been of any great significance, was cut off.

Some raised questions about this change, notably the United States. Actually, President Franklin Roosevelt had once questioned whether Yugoslavia should be reconstituted; his expressed opinion was that the "Croats and Serbs had nothing in common, and that it was ridiculous to try to force two such antagonistic peoples to live together." Serbia, he thought, should be established as a separate nation and Croatia should be put under a trusteeship. In the final analysis, however, the United States believed that the British knew more about the Balkans and that their lead should be followed. A damning source of information later was the more than five hundred Allied airmen, largely American, who had been rescued and protected by Chetnik forces. To the end they maintained their belief in the effectiveness and humanity of these forces.

The propaganda in favor of the Partisans was pervasive. For example, areas through which Tito's forces passed were often quickly listed as "liberated territory" even though they could not be held, and usually no effort was made to retake them. In London, Foreign Secretary Anthony Eden and some others had doubts, but for Churchill, seemingly, the only consideration was which side was killing the most Germans. When his trusted adviser Fitzroy Maclean came from Tito's headquarters to brief him in Cairo on his way home from the Teheran Conference and told him he believed the Partisans would gain power in Yugoslavia and establish a Communist system after the war, whether or not the West helped them, Churchill's response was to ask Maclean if he planned to live in Yugoslavia after the war. When Maclean answered no, Churchill said, "Neither do I, so the less that you and I worry about the type of political system that the Yugoslavs have after the war, the better."

At this time, the Partisans were principally entrenched in the mountains of northern Montenegro and Bosnia-Herzegovina. Some were in Slovenia, where they competed with a pro-Mihailović movement. There were almost none in Croatia. More important, there were few in Serbia. They appeared there in numbers only after a month's forced march following the Red Army's crossing into Yugoslavia. The Chetniks had been the first to welcome the Soviet troops. It was their tragedy. They were liquidated by the Russians and the Partisans. In Montenegro, the Partisans, who had been driven out earlier, were able to return to some areas after the capitulation of Italy in 1943.

Thus everything seemed to move with synchronization in Tito's favor. When the Italian surrender came, his troops were in a better geographic position than Mihailović's to seize Italian war matériel. Yet when the sizable Italian Venezia division in Montenegro, with its large cache of arms, surrendered, and Chetniks were in position to take advantage of it, they were, for some still unexplained reason, prevented from doing so by the head of the British military mission, who was with them. And, under the terms of the armistice, Italian commanders were obliged to surrender to Tito's forces.

The Italian capitulation not only meant more suitable bases for delivering aid to Tito, directed from Bari, on the Adriatic coast, but also indicated to the forces of Pavelić's Fascist Croatia that the Germans would sooner or later lose the war. Consequently, they began abandoning the sinking ship. But where could they go? The non-Communist Mihailović was far away, and there was also fear that his Serbian fighters might seek revenge for Ustashi massacres of their brethren. They therefore began joining the Partisans in ever-growing numbers, confident that their sins would be atoned for by assisting the Partisans to gain their ends.

Although no longer receiving aid, Mihailović continued limited actions against the Germans with the meager equipment at his disposal. Members of the U.S. mission to his forces were still assuring him that he would not lose out at the peace table, though they had been grievously wrong in assuring him that Soviet troops would not cross the Danube.

Such optimism on the part of the U.S. representatives was not warranted by other developments taking place. After strong pressure from Churchill, King Peter II, in May 1944, dismissed his cabinet. He made the former governor of Croatia, Ivan Šubašić, his new prime minister. The purpose of the change was to negotiate an agreement with Tito, which was done in short order. Tito and Šubašić agreed that Mihailović should be dismissed as commander-in-chief, but Peter was reluctant. Finally, when Churchill indicated publicly that Britain might be forced to proceed without his agreement, the king yielded, which was a deathblow to Mihailović's organization, particularly after the king was induced, in radio broadcasts to Yugoslavia, to urge his countrymen, including those with Mihailović, to join the Partisans in the struggle against the Nazis.

The Tito-Šubašić agreement was full of pro-democracy phrases and pledged that when the war was over the people would be free to decide their political future. It was also agreed that King Peter would be represented by a three-man regency pending a vote by the people on whether they wanted to retain the monarchy. Tito volunteered assurances that he had no intention of imposing on or introducing Communism in Yugoslavia. At a conference with Churchill in Italy, Tito personally assured the British prime minister of his position. He declared that he would allow individual freedom after the war, and added that "democracy and the freedom of the individual" were among their basic principles.

Tito and Šubašić also agreed that, pending elections, there would be a coalition government and a provisional parliament made up of Tito's Anti-Fascist Council of National Liberation and deputies from Yugoslavia's prewar parliament who had not been compromised during the war. Tito was designated prime minister, and Šubašić minister of foreign affairs. A few other ministers were non-Communist prewar politicians.

After the Soviet army captured the capital, Belgrade, with minor support from the Partisans, it was turned over to Tito's forces. Many years later, one of Tito's closest associates admitted that the Partisans could not have taken Belgrade alone, not even with the hundred or more tanks the Russians had given them. At another point, he reported that when they reached Belgrade they did not find a single Communist there.

Once established in Belgrade, Tito and his comrades threw all pretense aside. They introduced a reign of terror aimed not only against real or imagined collaborators, but also against the middle class and the intelligentsia, which was Tito's reenactment of Lenin's Red Terror in Russia a generation earlier. They openly demonstrated that the Tito-

Šubašić agreement was but another devious tactic in reaching their ultimate goal. Soviet-style elections soon proved that effective competition was not possible.

Non-Communists, including Šubašić, were pushed aside, and even when holding ministerial posts, they were not permitted to know what was going on in their own departments, much less to make decisions of any consequence. One by one they were driven by frustration to resign. Those like the independent socialist Dragoljub Jovanović who in prewar days viewed themselves as natural allies of the Communists attempted to stay on and do verbal battle with regime spokesmen, but they soon found themselves out of the government and sometimes in the dock or prison.

Šubašić had to go to a hospital. While there, he told his doctor about some of the things Tito and his comrades had done to force him out of the government. When the doctor urged him to put these things on paper, Šubašić answered: "No. I want to die in bed!"

In place of the liquidated coalition, Tito for a time sought to give the impression of a more widely representative government. Some third- and fourth-rate men from prewar parties were brought in, but they had no following and were not allowed to make speeches or hold party meetings. Their parties had to register with the authorities, though the only really active party — the Communist Party — chose to ignore the formality of registration.

The Tito-Šubašić agreement, which had received Big Three sanction at Yalta, was forgotten. The right of the people to choose their form of government was nullified by the spurious elections. The promised plebiscite on the future of the monarchy was never held. Instead, the Communist-controlled constituent assembly proclaimed the monarchy at an end. Before most people were fully aware of what was happening, a police state had been established. The dictatorship of the proletariat controlled everything.

The principal reasons for the success of the Communists have been touched upon. Nevertheless, it might be well to summarize them briefly.

First, from the beginning the Communists knew what they were fighting for. Everything was subordinated to the goal of a Communist state; no sacrifice was considered too great for its attainment; no means were rejected, no matter how repugnant, if they served to advance the desired goal. The Chetnik movement, on the other hand, was at a distinct disadvantage. The main purpose of Mihailović's Homeland Army was to assist the Allies in winning the war, not to determine what would happen afterward. Moreover, Mihailović's adherence to a semblance of a code of honor—as demonstrated by his concern for civilian populations subject to reprisals — and his respect for democratic methods left him the victim of a ruthless foe.

Second, the nucleus of the Partisan movement was the small but well-organized and disciplined Communist Party. The general staff of the party — its central committee — relayed decisions and orders through trusted party men in key positions. By always keeping the party in the background, and emphasizing the broader movement, it was possible to recruit unsuspecting patriots and thus to convey the impression that the movement was broadly based. Unlike the Communists, Mihailović had no previously developed organization, and was not in a position to exercise overall discipline. It was not difficult for Communists to infiltrate his movement and to cause difficulties within it.

Third, Mihailović believed that, as a resistance leader of a recognized Allied government, his task of providing help to the common cause would be shared by all patriotic citizens. He saw no need to publicize, or sell, himself or his movement. For the Communists, the common cause was but a convenient vehicle for reaching their ultimate objective. They devised a well-organized propaganda campaign for internal and external consumption. Even in the most difficult days, Tito's general staff was accompanied by party propaganda experts. Domestically, the propaganda was not always effective. In Serbia, for example, where the people knew that the Mihailović movement was truly patriotic, Communist propaganda had a hollow ring. For the foreign public, the propaganda stressed that the Partisan movement was broadly based and that its goals were democracy, freedom, justice, and a better future for all.

Fourth, the Partisans were aided by their geographically strategic location in the mountain ranges of Bosnia-Herzegovina and surrounding territory, mainly in Pavelić's puppet state of Croatia. It was to these areas that many Serbs fled to escape the Ustashi holocaust. They were eager to fight, and were mobilized by adroit Communist leaders, as were puppet Croatia's soldiers when it became evident the Axis would lose the war.

Fifth, the Partisans were helped by two external factors: Western aid and the Red Army. The Big Three decision at Teheran to make Tito's Partisans the chosen instrument of Allied policy in Yugoslavia opened the floodgates of aid from the West. Although Tito received virtually no material aid from the Soviets until late 1944, their propaganda and diplomatic activity was of immeasurable value. Later, the Red Army drove the Germans out of Belgrade and turned the capital over to his Partisans.

1945 Yugoslav Boundaries

7

The Tito Regime

Josip Broz (Tito) and, directly or indirectly, all his comrades learned their Marxism at the feet of Stalin and others in the Soviet Union. It was Marxism as modified by the leader of the Bolshevik revolution, Vladimir Lenin, and in Yugoslav and other Communist circles it was called Marxism-Leninism. This political theory or ideology has been the working body of knowledge or guidepost for all modern-day Communist parties. At the start, the Yugoslav Communists chose some leaders — Djilas, Kardelj, and Pijade — who were adept at quoting Marx, Engels, Lenin, and Stalin. Tito, who had lived for a time in the Soviet Union, was not himself a theorist, but, instead, a practical dogmatist.

In simplest terms, Karl Marx taught that the history of all hitherto existing societies was one of class struggles. The basic struggle is for control of the means of production, that is, the means by which society produces what it needs. Those who control the means of production will determine the nature of all political and other organizations in the society. In other words, the dominant class will establish the type of governmental organization that will protect its interests. In this way, the state becomes the instrument of the dominant class. When the means of production change, control of the new means will pass to a new class, and the nature of the political organization will change to fit the needs of this new dominant class. This is the dialectic: one thing creates a condition; another, opposition. The resolution follows: thesis, antithesis, synthesis.

Marx traced the historic changes in the means of production, and the political transformations that necessarily followed. Sometimes several classes had existed, but in the period in which he wrote, the latter part of the nineteenth century, he saw two opposing classes, the bourgeoisie, or property-owning class, and the propertyless proletariat, or working class. Capitalism, the economic system of the bourgeoisie, exploits the working class. When the latter grows strong enough, it will overthrow the capitalists. When it does so, the proletariat will be the dominant class, and will create a workers' state to replace the bourgeois state.

In fact, when the proletariat becomes the dominant class, it will not really be dominant, because there will not be any other class for it to dominate. Hence there will be no need for a state; society can move into the blissful condition of a classless and stateless utopia.

This desired condition does not, however, occur automatically. Marx argued that the proletariat must organize to overthrow capitalism. His followers, therefore, unlike other philosophers, not only seek to explain the world, but also show how to change it tactically. Marx also argued that there would be a transition period, during which a dictatorship of the proletariat would govern in the name of the working class. Lenin, impatient to witness the arrival of the revolutionary condition in Russia, concluded that the transition could be hastened through a small tightly organized and disciplined party of professional revolutionaries.

From this flowed the notion of a leader — for example, Stalin, Tito, Castro, Mao — whose word was, in effect, law. Any questioning of his decisions, even by party insiders, was tantamount to treason and usually punished as such. Evidence of this is to be found in the vast and bloody purges of the Soviet Communist Party in the 1930s. The brutality of Stalin's regime has been attested in many ways and by countless witnesses. Milovan Djilas, the Yugoslav Communist who after 1954 became a dissenter and was often imprisoned, has repeatedly declared that Stalin killed more Soviet citizens than Hitler had killed German citizens.

Communist Party purges of the populace have occurred in other countries, including Yugoslavia. Tito and his comrades proceeded to copy the Soviet system lock, stock, and barrel. As in the Soviet Union, there were proclamations that the new system constituted the most advanced type of democracy, that the country was now a "people's democracy." In reality it was a political system in which the Communist Party was the driving force, a dictatorship not really of the proletariat but over the proletariat and all other groups in Yugoslav society. No competing groups or influences were tolerated. The party did not stop with governmental affairs. Every phase of human endeavor was within the scope of its all-embracing authority.

The leadership sought to justify the party's position by claiming that it was, as was the entire political system, democratically operated. The party rank and file elected delegates to a party congress, which chose a central committee. The central committee chose a politburo as the party's supreme authority. The most powerful person in the politburo was the general secretary. Since all these party bodies were really "democratically" elected, those who elected them had an obligation to obey the orders of those whom they elected. This is what is known in party jargon as democratic centralism. Party members have, at appropriate intervals. an opportunity to choose a new leadership, and in theory they retain the right to criticize the work of party officials and party governing bodies.

In actuality, those who reach the top, by whatever means (Tito was sent by Stalin to be general secretary of the Yugoslav Party), cannot be challenged by those below except at the risk of ouster or even death. Nominations and elections of party officials are dictated from above, through party organizations. At a party congress, for example, the delegates are handed lists of nominees for the central committee, which are to be dutifully elected. Similarly, the politburo presents to the central committee the names of persons it wishes dismissed or added. In sum, the party structure becomes a self-perpetuating hierarchy.

In terms of functional control, the general secretary is assisted by a secretariat. Within this body, which usually has several politburo members, there is some division of labor. One member may be the party's specialist in agriculture; others may be in charge of party affairs, agitation and propaganda, security, or foreign affairs, and so on.

The governmental structure is similar; it, too, is a hierarchy, a pyramid of power. Party members are dominant, and at or near the top they are *all* party members. In lower positions are many non-party people, which helps to portray the government as not run solely by the party. Non-party persons are, of course, carefully screened by the party, to assure complete loyalty. The government, therefore, can be seen as constituting an administrative apparatus for implementing party policies.

One of the first tasks Tito and his comrades confronted once they were in power was to reorganize the state geographically. They had criticized the alleged dominance of Serbs in prewar Yugoslavia, though they were, after all, the largest ethnic group, almost half of the country's total population. Additionally, while seeking support for the Partisan movement, they had stressed the need to give all nationalities a fair break, and had campaigned under the slogan "Brotherhood and Unity." They had had little success in Serbia, which had overwhelmingly supported the resistance movement led by Mihailović. Now Tito determined to punish Serbia. It has been alleged that he was motivated by his own parentage — his father was a Croat, his mother a Slovene — but this is doubtful. His motivation was primarily to secure his own power.

The country was given a Soviet-type federalism, which in point of fact did not have the essential attributes of federalism, understandably so in view of the fact that the Yugoslav state was a dictatorship. The country, under so-called federalism, was divided into six republics — Serbia, Croatia, Slovenia, Bosnia-Herzegovina, Macedonia, and Montenegro. As a way of weakening Serbia, many Serbs were left in each of the other republics except Slovenia. The largest number, more than a million and a half, were included in Bosnia-Herzegovina, which was created as a separate republic, partly because the large Muslim population made up forty-five percent of the total. These latter were Serbs and Croats who in centuries past converted to Islam under pressure or threat or for the promised favoritism offered by their Ottoman Turkish rulers. The Serbs accounted for about thirty-four percent of the total; Croats, about seventeen percent.

Not until the 1971 census, however, were inhabitants of Bosnia-Herzegovina encouraged to declare themselves ethnically Muslim, the only Yugoslav republic where this has been standard practice. It is ironic that among many supposed "original" ideas of Tito's system was the one of equating ethnic origin with religious affiliation, and this occurred in an avowedly atheistic Marxist state.

The republic of Montenegro, although it makes up about four percent of the Yugoslav population and is almost entirely Serb, was made a separate unit in large part because so many of Tito's comrades came from there and wanted to have a separate republic. Then, too, the Montenegrins are fierce in war and politics, and Tito was aware of their proclivities. As for Macedonia, it was made a republic principally to deny Bulgaria a propaganda weapon. The Communists in Yugoslavia wanted no more 1913 squabbles with Bulgarians.

As a further insult, Serbia was forced to accept two autonomous provinces on its territory, Kosovo-Metohija (usually just Kosovo) and Vojvodina. The insult was grave as regards the first, because Kosovo is the cradle of the Serbian nation and the Serbian Empire ended there on Vidovdan in 1389. In the province established by Tito, Albanian Muslims soon outnumbered Serbs. In Vojvodina, the Serbs were in a majority, but there was a sizable Hungarian minority and not insignificant numbers of Romanians, Slovaks, and Ruthenians.

When questions of autonomy were being discussed, one of Tito's comrades, Moša Pijade, recommended that Dalmatia, which prior to the creation of Yugoslavia was never a part of Croatia, be given autonomy. Tito cut off the debate with "Don't split up my Croatia."

By this organization, the regime insisted that the nationality problem had been solved, and that Yugoslav unity had been achieved. What they had done, of course, was proclaim the problem solved, and they imposed stiff penalties on anyone who suggested otherwise. In actuality, they had aggravated the problem by making it impossible to discuss unresolved issues. Perhaps the Macedonians felt some satisfaction in having gained recognition as a separate nationality group, but the Serbs living in Macedonia (to them, the old South Serbia) did not like having to change their names (for example, Jovanović to Jovanovski). The Muslims in Bosnia-Herzegovina also may have been gratified to have their identity recognized. Many Serbs living there, however, wanted to retain their Serbian identity even when they were of the Islamic faith.

The actions of the Yugoslav Communists in trying to resolve the nationality question were in sharp variance to the stand of the party in the interwar years. The official line of the Comintern, directed by Moscow, had been that Yugoslavia was an artificial creation of the bourgeois circles that had dominated peacemaking at Versailles and should be destroyed. The Comintern had harangued against the Serbian bourgeoisie, insisting that it was exploiting the other nationalities of Yugoslavia. The specific rights of Croatians, Slovenes, Montenegrins, Albanians, and others to their own states were expounded in declarations that Yugoslav Communists generally endorsed in public. Privately, however, no question divided Yugoslav Communists quite as much as the nationality question. More than one general secretary of the Yugoslav party perished in the Soviet Union during the interwar years simply because they dared to point out to Stalin that the Yugoslav situation was far from simple.

Some of Tito's decisions aggravated the nationality problem in at least two ways. First, the regime adopted a policy followed by prewar governments of investing in industrial enterprises largely in Croatia and Slovenia, which were the most developed. This held back the less well-developed republics, which had been promised development aid but got less than anticipated. Second, the regime sent officials of other nationalities to various republics, a practice the Communists had criticized when it was done by prewar regimes. In Croatia, for example, Serbs now seemed to be in control of the national government. Actually, those sent to Croatia were loyal Partisans from Serbian areas in Croatia or from Bosnia-Herzegovina, and they did not have much power. Real power resided in the hands of leading Croatian Communists. In Serbia, the presence of many non-Serbs led Serbs to conclude that Croats and others were in charge of things. Yet, everywhere, real power was in the hands of Communists trusted by Tito.

The governmental institutions that were established had a superficial resemblance to democratic systems in the West. There were legislative bodies elected by the people at both national and local levels. At the national level there was a legislature of two houses, one to represent the people generally, the other to represent the republics. At the republic and local levels the legislatures were unicameral. These bodies were, as in the Soviet Union, actually administrative instruments for carrying out party directives. Elections were party controlled. Party organizations determined the nominations, and only one name appeared for each office to be filled. There were no opposition lists or alternative programs from which the voter could choose. It was not the purpose of elections, as some leading Yugoslav Communists admitted, to choose those who would govern. One of Tito's close associates, Milovan Djilas, readily admitted in the election campaign of 1950 that the purpose of the election was "to strengthen the dictatorship of the proletariat" and to confirm and explain the party line. Moreover, he said, "the question of whether the candidates will be elected is of secondary importance, because under such conditions and under such a system they certainly will be."

Hence, the purpose of elections was to give the party another opportunity to present its message, in a different context. Perhaps such elections provided the leaders with the feeling that they had popular support. The impact on the voter, of course, was a certain amount of frustration or humiliation for having voted for something he or she did not believe in.

In the judicial sphere the Yugoslav Communists also imitated the Soviets, by destroying the existing legal and judicial machinery and replacing it with laws and courts like those in the USSR. One difficulty was the lack of politically acceptable persons with legal training. Since the party could be relied on for instructions and guidance, this was not a handicap in cases important to the state. A curious aspect of dictatorial rule in this area was the serious, detailed consideration of what was "legal" and "properly procedural."

Bureaucracy tends to take seriously its own rules, whether or not they are popularly created and confirmed.

In the years after Tito's break with Moscow, in 1948 (see next chapter), a great deal of experimentation took place in both political and economic spheres. The basic motivation was the conviction that Yugoslavia must show that it was different from Moscow's or other Communist systems. Many of the attempts in experimentation were based on trial and error and doomed to costly failure.

Because they had gained power by force, rather than through free elections, the Yugoslav Communists, like other revolutionary regimes, needed to make sure their power was secure. They felt the need for this especially because their regime was not popular among a large segment of the public. They had learned this not only from experiences during the war, particularly in Serbia, but also from the meager vote they had received during the First Yugoslavia whenever they had candidates. They therefore quite naturally employed the same tactics used by their mentors in Moscow: a combination of repression and persuasion. A secret police force, a replica of the Soviet model, was established even before Tito assumed full power. And a series of steps was taken to gain complete control of the media, and therefore of public opinion.

Their use of force and fear required first of all the physical elimination of those judged to be a threat to the regime. Many victims were identified before the Titoists took power. Those considered the greatest threat — politicians known for their anti-Communism, judges who had sent Communists to jail, wealthy businessmen, bankers, and teachers, lawyers, and journalists known for their anti-Communist views or actions — were rounded up and quickly executed, often without even a perfunctory trial. Those considered less dangerous, and often not necessarily associated with the bourgeois class, but a significant probable threat were sent to prison for varying terms. They generally received trials, but of the Soviet type, in which defense was largely futile. A third group, not sufficiently dangerous to require incarceration or liquidation, was placed under surveillance by informers recruited by the secret police.

On what might be called the positive or persuasive side, to harness the masses to the dictatorship, the party spewed forth a plethora of organizations, stemming from the central root, spreading in all directions, dividing and subdividing. Nearly all were parasitic — economically unproductive, their budgets eating into the state treasury. Ambitious to expand, and consequently always in need of additional resources, they published newspapers and magazines, organized special outings, and held congresses.

The most notable of these mass organizations were the People's Front, the People's Youth, and the Labor Syndicate. Among those of more limited import were looser associations designed to encompass actors, teachers, government employees, students, writers, journalists, or athletes. Many persons belonged to more than one of these.

These organizations produced a captive audience to whom the regime's propaganda could be served in liberal doses. It was hoped, too, that people so engaged would have little time to think about the injustices of the regime.

The mass organizations fundamentally engaged in two types of activity: They performed some useful physical work in the postwar reconstruction period, and, more important from the regime's point of view, they served as vehicles through which the party could sell itself and at the same time agitate for greater effort in the development of socialism. Much pressure was exerted to get members to volunteer. In the case of university students, for example, so-called voluntary work was the price of being permitted to continue their studies.

In an effort to produce youthful enthusiasm for the new order, propagandists devoted a great deal of time to publicizing so-called voluntary youth work projects. Where it was necessary to impress foreign visitors, special squads were provided and were instructed in how to make a good impression. As the years passed, however, many such projects fell by the wayside.

Yugoslav Communists devoted more attention to the young than to any other segment of the population. Their importance to the future was evident in the assignment of party members, under twenty-four years of age, to work with them. Before adolescence, children were made members of organizations whose basic aim was to condition them to be "future builders and governors of their socialist homeland." As they moved into more advanced groups, the youngsters were to be educated "along the line of our revolutionary and heroic past."

The schools were naturally felt to be the most effective instrument in shaping youth and producing future party leaders. The task would mean reshaping a people's mode of thought and behavior, revising their national, cultural, and spiritual values. To accomplish this, the leadership turned again to the Soviet model. They established fundamental ideological goals and created the bureaucratic machinery to achieve them. Subsequently, they were to depart from that model, but they continued to believe that it was the function of the central authorities to determine educational policies.

Perhaps their principal immediate problem was the teachers. From the party's point of view, they were unreliable. Yet no others were available. Consequently, a major effort was made to devise special ideological courses for them, after which student informers would be utilized to report on their conduct in the classroom. The party had just begun the process of getting newly trained teachers when the break with Stalin came, making it necessary to "reeducate" pro-Soviet Communist teachers. The lack of significant success was voiced by Tito in the early 1950s when he declared that many of the teachers in the schools were "total strangers to our reality," and in the universities, he continued, the "situation is still worse."

Another aspect of the regime's concern for the young was the perceived negative influence of home and church. To counter the influence of the family, special weekend activities were devised for youngsters at an early age, and clever entertainment was provided on Sundays, to compete with churchgoing. In schools, religion was ridiculed, as were children known to go to church. At the same time, Yugoslav leaders early recognized that indoctrinating the youth was not easy. Courses in theories of Marxism-Leninism were frightfully boring. And countering the influence of anti-Communist forces, family and church in particular, was a difficult task.

The regime devoted a great deal of effort to countering the influence of religion. Rigid laws were enacted limiting what church authorities could do, particularly in seeking converts and soliciting money. Permission to erect new churches was usually denied. In some Muslim areas, women were compelled to discard the veil by being threatened with arrest. Pro-regime associations were created within each of the major religious denominations, to show the outside world that the government had support among religious groups.

Party functionaries made sure that members of all groups, including the pro-regime clerical associations, used the media to voice their satisfaction with the new society. If a group did not have a propaganda outlet, the regime created one. Consequently, if a person was asked to write about the pleasures of his or her profession in a socialist society, it was difficult to refuse on grounds of not wanting to get involved in politics, because the request was not to write for a political publication, but for one in the person's own field. If such a person claimed to have no talent for writing, the party was glad to provide a ready-made text.

For the general public, the regime controlled all media outlets. There were no privately owned newspapers or privately owned or operated radio stations, no privately owned or privately operated publishing enterprises, no privately owned theaters or movie houses, or privately produced programs for them. The party's specialists in agitprop (agitation and propaganda) provided guidance to editors and all others whose work would be presented to the public in whatever form. This work was headed by the party's ideological specialist in the politburo.

The totalitarian nature of the regime prevented any manifestation of opposition. Yet there was one notable exception. This occurred at the University of Belgrade in 1968. So-called student demonstrations originated in what the students regarded as unsatisfactory housing and other facilities. What was significant, however, was that they used the opportunity to widen the area of complaint. Basically, they complained about the contrast between the poverty of student life and the opulence of life for the ruling class. In the student newspaper and elsewhere, among the most telling slogans were: "The wheel of the Mercedes is not the wheel of history," "Down with the Red Bourgeoisie," "Down with the princes of socialism," "Freedom, Truth, Justice."

The regime responded with repression, which brought forth more student demands, as well as condemnation of police brutality. To resolve the issue, Tito, on radio and television, admitted that in the case of certain domestic problems the government had not moved fast enough. He promised action and asserted that he would deal with those responsible for brutality. If he could not deal with the problems, he would no longer remain in office. He urged them to go back to their classes and exams in the meantime. Nothing much came of his promises. But the demonstrations showed that even the young people on whom the leadership placed such great hopes had serious misgivings about the new order.

In the economic sector also, Tito and his comrades, true to form, turned to Soviet experience for guidance. Both societies were primarily agricultural, in Yugoslavia's case eighty-five percent. According to Marxist theory, the proletarian revolution would come in the later phases of capitalism, when there was a highly developed industrial society, with concentration of wealth and a large exploited class-conscious proletariat. Yugoslav society was not industrially developed, nor did it have a wage-earning class of any consequence. But to the Yugoslav Communists theory was not as important as Soviet practice. Moreover, the Soviets had elaborated doctrines to justify Communist revolutions where the economy was primarily agricultural.

In the first years of their rule, the Yugoslav Communists wanted to profit from Soviet experience in seeking a shorter route to a functioning economy. The lack of an industrial base, the shortage of skilled workers, and the virtual non-existence of managerial know-how did not dampen their determination to build an industrial society on the Soviet model. Nor did they wait for the establishment of a new constitutional order that would endow them with appropriate legal authority before plunging ahead.

Their efforts resulted only in disruption, chaos, and confusion. Yet they always displayed an assurance of knowing where they were going, of having no doubts about their ability to get there. Their answer to repeated failures was to change, improvise, try anew. This involved a steady stream of directives, decrees, and instructions. Costly mistakes and intolerable waste did not seem to be of grave concern to Stalin's imitators.

While they were still seeking to gain power during World War II, and assuring the Allies that private property would be respected in the postwar period, Tito's Partisans were busy confiscating industrial enterprises. By the end of the war, they admitted, more than half of the country's industry was in their hands, chiefly through confiscation. This was justified on the alleged grounds that the owners of these properties had collaborated with the enemy.

Ratifying for the most part what had already been accomplished was a sweeping nationalization law in 1946. Other laws were quickly enacted to take care of unforeseen circumstances. A supplementary nationalization law in the spring of 1948 brought almost every remaining economic enterprise, excluding landholdings, under government ownership. This included resort hotels, the few remaining restaurants, garages, secondhand bookstores, and other small-business establishments. Unlike the 1946 law, the later one came without any advance warning and went into effect on passage, not after publication in the official gazette. The next morning government agents were at the doors of establishments to be nationalized. Owners were not permitted to take anything. Even the money in their safes and cash registers was nationalized.

Although both laws provided for compensation, owners were merely given a slip of paper as a receipt. Years later they received only a fraction of the real worth of their properties, and that in highly inflated currency and government bonds that turned out to have little value.

Property that was not nationalized, such as private landholdings and residences, came under rigid control. Prices were fixed on most services and commodities, including rents. Full power to allocate housing space was assumed by the government, even in private homes whose owners did not desire to have tenants. Transfer of real estate was forbidden, even as a gift, without prior authorization.

Several months after the first nationalization law, the regime launched a five-year economic plan that sought to establish a collectivist economy. Following in the footsteps of Soviet planners, the Yugoslav Planning Commission placed major emphasis on industrialization and electrification. At the same time, the country was to cut traditional economic ties to the West and reorient itself with Eastern Europe.

In the agricultural sphere, Yugoslavia was for the most part characterized by small landholdings. Marxist theory had taught the Communists that the peasant was backward and conservative; yet more than fifty percent of the land belonged to poor or near-poor peasants who owned fewer than twenty-five acres. The Communists did not let these facts interfere with their determination to collectivize agriculture. They set in motion a whole series of measures and implemented them with ruthlessness. As a first step, they confiscated many acres of fertile land that had belonged to the German minority of about 500,000. This was justified by charges that they had collaborated with the Nazis. Some had, indeed, fled with the German armies; others were "resettled," presumably somewhere in the Soviet Union. For other lands confiscated from alleged collaborators, the charges were often undocumented and sometimes no effort was made to prove them. In addition, a Communist agrarian law relieved churches and absentee owners of their lands.

On some lands acquired by confiscation and the agrarian reform law, the government established farms patterned on state farms in the Soviet Union. In distributing others, first priority went to deserving members of Tito's Partisan army, their orphans, or landless peasants who had contributed to Communist successes. Most of the last came from mountainous regions where arable land was scarce or of poor quality. As in the Soviet Union, these peasants became wage earners.

Simultaneous with these moves was the establishment of

a system of compulsory deliveries, assessed against all peasants who still owned land. They were obliged to sell their surplus produce to the government at set low prices, and, worse, government agents determined the surplus, which often exceeded the peasant's total crop. This was an effective way of punishing better-off peasants, who sometimes had to purchase produce from poorer peasants at high prices and then turn it over to the government at a fraction of the cost. In many cases, a peasant was forced to sell a cow or other asset to acquire the money to buy the required produce.

In the years after the break with Moscow, Tito and his comrades made many changes, in both the political and the economic systems, and in related areas. The fundamental point is that these were controlled changes, with one factor remaining constant: The Communist Party's position and Tito's personal authority were not to be threatened.

The break with Moscow can in some sense be said to have "saved" Tito's Yugoslavia from some of the continuing hardships elsewhere in the Soviet orbit, in Romania and Poland and Bulgaria, as examples. Once the West saw that Stalin could be challenged, it began to regard Yugoslavia favorably. Tito had taken the chance of challenging Moscow's internal interference because, mainly, of Yugoslavia's geographical position. He told Djilas, "The Americans will never allow the Soviet Union to control the Adriatic." He was right, and Stalin knew the risk of invading Yugoslavia in 1948 while the United States was still heavily armed and ready to confront crisis, as it did with the Berlin airlift. The point about the Adriatic is reinforced by the Truman Doctrine, adopted in 1947 when the British told Washington they could no longer supply and support non-Communist forces in Greece. President Harry Truman, in effect, declared Greece to be under United States protection. The significance of Tito's challenge to Stalin is borne out further by the credibility Western liberals began to invest in "official" Yugoslav pronouncements and policies.

In general, it can be said that Yugoslavia continued after 1948 to suffer setbacks and failures in its civil affairs, economic and political. For a time, however, it appeared to be a great deal freer and more prosperous than other Communist-ruled states in Eastern Europe. Material prosperity was mostly evident in the late 1960s and early 1970s, but a notable decline had set in by the 1980s.

Politically, an early act was to change the name of the Yugoslav Communist Party in 1952 to the League of Yugoslav Communists (LCY), but this did not really represent any significant modification of the party's role in society. In addition, electoral laws were modified so that it was possible for more than one candidate's name to appear on the ballot for an office. The structure of the national parliament was changed more than once. At one time, membership in the so-called upper house was based on producers. The so-called lower house was, for a while, to represent the people directly and in proportion to population; so larger republics had more deputies than the smaller ones. A subsequent change gave an equal number to each republic; so a republic of several million had the same number of deputies as one with considerably fewer than one million. This was viewed as unfair, particularly in the larger republics.

The greatest political change was made in 1974 when the party's theoretician, and at that time perhaps Tito's closest associate, Edward Kardelj, a Slovene, produced a new constitution. It was eighty-four printed pages, without doubt the world's longest. The system it set forth was complicated beyond belief. Many of the party's leaders, let alone the average citizen, found it difficult to fathom.

This constitution, for all practical purposes, required unanimity among the republics and the two Serbian autonomous provinces for the making of all important decisions. In critical or emergency situations, the country's collective presidency could make decisions if the parliament and the cabinet could not agree. The eight-man collective presidency, consisting of one from each of the six republics and the two autonomous provinces, was the handiwork of Tito. He was president for life, and clearly in a category by himself. He had realized that he might not have too much longer to live, and he reasoned that a collective presidency would prevent a struggle for leadership among his party comrades.

The result, following Tito's death in 1980, was almost complete political paralysis. To avoid deadlock, the leaders often cast their decisions in the most abstract form, so that each of the members of the collective presidency could interpret it to his constituents in any way he chose. Understandably, in many instances confusing outcomes were the result. In the time of crisis in 1988–90, for instance, Tito's revenge on Serbia after the war served to thwart the separatist Slovenes and Croats. Serbia and Montenegro were joined by the two autonomous provinces Tito had carved out of Serbia, Vojvodina and Kosovo; their four votes prevented the other republics from gaining a majority. Tito did not foresee the irony in his handiwork, i.e. that his system would produce paralysis.

In the 1960s and 1970s, the political atmosphere allowed for greater creative and personal freedom. The dead hand of conformity that Stalinist "socialist realism" had forced artists and writers to labor under gradually gave way and party interference became less and less noticeable. For the average citizen life became more relaxed, and it was possible to travel abroad if one had the means.

In the economic area, the most dramatic change came in agriculture, where great dissatisfaction became more and more ominous. Faced with a rapidly deteriorating state of affairs, the Yugoslav leaders in early 1953 issued a decree permitting the reorganization or liquidation of labor cooperatives. Yet, other provisions discouraged peasants from leaving the collective: They would not be allowed to keep land they had contributed; they must continue to assume part of the collective's debt. In the face of a party declaration that the aim was to strengthen collectivism, rather than weaken it, the response was immediate: the vast majority of peasants again became private owners of land.

In industry, the changes were less dramatic, but useful nonetheless. Central planning was largely limited to setting long-range goals. Some attributes of a market-divested economic system were introduced, but under controlled conditions. Competition among government-owned enterprises was permitted, though this led to some unexpected and undesired results. Among these were the ruinous practice of undercutting prices, and duplication — so that the manufacture of stoves, for example, was excessive because it was to be found in several republics.

Even as some form of market economics was sought, the management of enterprises was altered to reflect more accurately Marx's vision of workers being in charge. To that end, workers' councils were created. In practice, the party was able to direct things through the manager; the workers could do a great deal of discussing, but had only a minor voice in making decisions. Subsequently, the concept of self-management was introduced, and had a considerable influence abroad. In practice, party control was, by and large, maintained, partly because the rules were so complex they were not easily understood by the individuals chosen by the workers to implement the system. After a number of years, the system was generally considered to have failed of its objectives.

In the 1980s, following a rapid and steady fall in the standard of living, Yugoslavia had difficulty meeting international loan payments, which needed constant restructuring. These loans approximated twenty-three million dollars. Many Yugoslav citizens were unaware of how much indebtedness they had been saddled with by Tito, who was no longer around to help them with the burden.

8

The Tito-Stalin Break

In June 1948, the world was electrified by the news of a schism in the camp of international Communism. Yugoslavia's expulsion from membership in the newly created Communist Information Bureau, the Cominform, broke the solidarity of the Soviet bloc of nations. Prior to the break, no one thought this possible. Soviet theoretician Nikolai Bukharin had declared in 1936 that rivalry between Communist states was "by definition an impossibility." He pointed out that the capitalist world was "made up of selfish and competing national units and therefore is by definition a world at war. Communist society will be made up of unselfish and harmonious units and therefore will be by definition a world at peace."

At the time of Bukharin's statement, there was only one Communist state, the Soviet Union. Following the creation of several other Communist-ruled states after World War II, the situation changed. The old Communist International, the Comintern, which was totally controlled by Moscow, had been abolished in 1943, perhaps as a way of smoothing relations with the Western Allies in the midst of war. In 1947, the Cominform was created, presumably for the purpose of exchanging information among the Communist parties, though Stalin seems to have viewed it as an instrument to deliver directives from Moscow.

Belgrade was made the headquarters of the Cominform. Some maintain that this decision was based on Stalin's supposed trust in Tito. Others insist that it was precisely because Moscow did not trust the Yugoslav dictator. In either case, by locating the Cominform at the center of Tito's power, Stalin could have his agents at hand as a way of controlling what happened in Yugoslavia.

It should be noted that in the prewar period, Stalin had experienced problems with Yugoslav Communists, and several general secretaries of the Yugoslav party perished in his purges of the 1930s. In 1937, he sent Josip Broz, under the name Walter, to be general secretary of the Yugoslav party. Obviously, he had confidence in Tito, and it had seemed to the outside world that he had not made a mistake. Domestically, Tito was organizing the country on the Soviet model, and in international forums he was a vociferous defender of the Soviet Union and a severe critic of the United States.

Largely unknown to the West was the fact that Stalin had had problems with Tito during the war. The Soviet Union had sowed seeds of distrust by failing to send material aid to the Partisans, although there were words of encouragement for their activities. Stalin was critical of Tito particularly for organizing his guerrilla movement in his own way, without Moscow's advice and direction. More specifically, Tito and his comrades were criticized for setting up proletarian brigades, for using red stars on Partisan caps, and for establishing people's committees as governing bodies in liberated territory. Stalin apparently was worried that these acts would create difficulties for him with his Western allies, who at that time were supporting Mihailović's resistance group. Tito rejected the criticism and other forms of advice, and thereby demonstrated his independence.

It is interesting to remember that one of the first things the Soviets attempted after their troops made contact with Tito's forces in Yugoslav territory in the autumn of 1944 was to put Tito's units under the direct command of the Red Army. Tito refused. The Russians were not easily dissuaded and sought to have their way several times, but Tito held firm.

Tito had other difficulties with the Red Army, whose soldiers had looted and raped. One of his commanders voiced his displeasure with the behavior of Russian troops, and this was reported to Stalin. These quarrels with the Red Army, as well as the other wartime difficulties, might have been surmountable. Instead, they constituted a harbinger of more far-reaching and more disturbing postwar troubles.

It became clear to Tito and his comrades that the Soviets wanted to dominate Yugoslavia's domestic and foreign policy. They wanted to determine the course of the country's economic development. Yugoslavia was to remain an exporter of raw materials and agricultural products, for which the Soviets would dictate markets and prices. Moreover, they wanted freedom to explore for raw materials inside Yugoslavia. They also expected the Yugoslavs to accept what was sent to them, even articles they did not need, and pay the prices asked. The Yugoslavs were forced also to conclude commercial agreements that allowed the USSR to procure goods from Yugoslavia at world market prices, which were many times lower than the actual cost of producing them by the semiprimitive methods still existing in Yugoslavia. In short, the Soviet Union expected to engage in economic exploitation far more invidious than that of which it had accused the capitalist world.

Aggravating all this was the fact that goods and services from Moscow had to be paid for in gold or hard currency. Still worse, some of the military equipment, paid for in gold, was found to be old and defective, although newly painted. Gunpowder was delivered in boxes containing rejection slips.

The Cominform resolution that expelled Yugoslavia from membership pointed out that Tito and his associates had been "invited" to appear at a meeting of all the members and to answer questions about their heresy. The Yugoslavs had refused to do this. The resolution depicted them as conceited and immodest, unwilling to admit their mistakes and to engage in self-criticism.

The resolution heavily accented ideological issues. Among other things, the Yugoslav Communists were charged with relinquishing their leadership role by burying themselves in the Popular Front, taking the road of nationalism, and failing to collectivize agriculture. The "Tito clique" had betrayed Marxism and sold out to the imperialist West. There was some logic to this charge, because the Soviets had set themselves up as the final arbiters of what was or was not Marxism. It mattered little that Tito or others might disagree; that very disagreement constituted proof of heresy.

To accept the type of economic domination envisaged by Moscow, the Yugoslav Communist party would have to become completely servile to Moscow. In other words, political domination would have to go hand in hand with economic domination. According to Djilas, friction first arose between the secret services and the propaganda establishments of the two countries. Soviet recruiting of Yugoslavs for their own intelligence service, he reports, was systematic and aggressive. The Soviets also recruited agents and informers among the Russians who had gone to Yugoslavia after the Bolshevik seizure of power and the children of those immigrants born in Yugoslavia.

Soviet agents, in the guise of military or economic specialists, penetrated various branches of the Yugoslav government. They worked in the Communist Party as well, endeavoring to build an anti-Tito wing. It is unclear when they began to undermine Tito's leadership, but it was at least by early 1946. Fortunately for Tito, his agents were alert and succeeded in frustrating Moscow's designs. These Soviet agents and their Yugoslav assistants were rounded up before they could do any damage.

Although less specifically spelled out, Stalin seems also to have been more than a little concerned about Tito's independent acts in other Balkan countries. He viewed this as encroachment on his authority; he did not want another Communist center outside his purview. He may even have tolerated this as long as Tito was working for Moscow, but when he realized that Tito was acting completely as an independent entity, he had to go. The Yugoslav leaders acted as an independent power, while the Soviet Union acted as an expansionist world power.

Stalin overestimated his ability to deal with Tito. In the past he had been able to purge leaders of Communist parties with relative ease. His first move against the Tito clique (primarily Tito and three close associates) was to call upon the "healthy" elements in the Yugoslav party to compel the leaders to admit their mistakes and to correct them. His confidence can perhaps be best characterized by a comment to his associates: "I will shake my little finger and there will be no more Tito. He will fall." No doubt of major importance, which Stalin may have forgotten, was that the leaders he had purged in the past did not have a power base. In no other Comintern country had there been so large a military resistance force as in Yugoslavia. Tito had emerged from the war with a Communist network in much of Yugoslavia, a unique base for later action. In addition, Tito could rely on a group of younger subordinates who owed their rise to prominence to him alone and who had never had links with Moscow.

Stalin's next move, probably too late, was to rely on help from pro-Soviet members of the politburo. But Tito arrested two of them, Croatian Andrija Hebrang and Serbian Sretan Žujović, who he felt certain were sympathetic to Stalin, as well as some lesser figures. He became merciless toward suspected Stalinists. About 12,000 were rounded up and sent to an island in the northern Adriatic known as Goli Otok (Naked Island), where they were subjected to brutal treatment. A few Cominformists escaped or were shot while seeking to flee.

Another aspect of Stalin's campaign against the Tito clique was psychological warfare. Pro-Soviet organizations sprang up in all the East European states, among them a new Yugoslav Communist Party. They published newspapers, pamphlets, and even engaged in terrorist border incursions. Threatening troop movements were observed along Yugoslavia's boundaries. None of these actions seems to have had any significant effect.

There were purge trials of Communist leaders in Prague, Budapest, Sofia, Warsaw, Bucharest, and Tirana to prevent defections similar to Tito's. These were organized in Moscow, and, as was usual in such trials, most of the accused confessed to their errors. In fact, the "Titoist disease" did not spread.

Perhaps most important was Stalin's effort at an economic blockade. After the Communists came to power, Yugoslavian trade had been oriented to the Soviet Union and the other East European states. Now all of them reneged on their trade commitments and canceled treaties of friendship and mutual assistance. It is of interest that these steps were not taken until about a year after the Stalin-Tito break, perhaps because only then did the Soviets realize that their initial measures were not succeeding. Soon the Cominform adopted a resolution condemning Tito as a Fascist spy in the pay of the imperialist Western bloc. Not until late 1949 was there mutual expulsion of diplomats, although there were no breaks in diplomatic relations.

There is some evidence that a final effort to oust Tito was a planned military invasion, to be orchestrated from Hungary. That it never took place seems to have been due to wider considerations. First, Tito's early belief that the Adriatic was Yugoslavia's safeguard—that the United States would not allow the Soviet Union to face the West there and control Greece—was still valid. Second, a purge of Hungary's strategic leadership, charged with being Tito's puppets, included General Bela K. Kiraly, who had been designated the commander of the planned invasion. Third, the United Nations action in Korea may have convinced Stalin that a similar action might occur with respect to Yugoslavia. Both

U.S. and British foreign secretaries had earlier issued public statements putting Moscow on notice that an attack on Yugoslavia would have serious consequences.

There are a number of reasons Tito was able to withstand the efforts of Stalin to overthrow him. First of all, those around Tito knew that he had been a faithful Communist, and Communist rule in Yugoslavia was not threatened. With the exception of those purged by Tito, there was considerable continuity in the leadership of the party. Moreover, party rank and file were made aware of the anti-Yugoslav actions of Soviet authorities.

Another reason was the financial and moral support provided by the United States. Unhappy about the oppressive nature of Tito's regime, the United States, convinced that the split with Moscow was real, found it the lesser of two evils. One of the first U.S. actions was to unfreeze some $30 million in gold that had been deposited in New York by the Yugoslav government on the eve of World War II.

The United States hoped that in the long run the Communist and totalitarian Yugoslav regime would move in the direction of democracy. In the period immediately after the break with Moscow, there was no appreciable rise in Tito's popularity among Yugoslav citizens, who had hoped that Tito's defection might be the beginning of the fall of Communism, but this was not to be.

Another element in Tito's favor was that in 1948 the Soviet Union had its own domestic and international problems. The war had seriously weakened the nation, and American aid to the Soviets ceased after President Truman realized that the Soviets were not going to live up to Allied agreements concerning Eastern Europe. Earlier, the United

States had sent aid to Greece and Turkey, and the Marshall Plan, to ensure recovery in Western Europe, was launched prior to the Stalin-Tito split. Tito had been supplying Greek Communist guerrillas, while Britain and the United States assisted the established Greek government. Consequently, by 1948 Moscow could see evidence that the United States and Western Europe were determined to resist Soviet expansionism. The Soviet Union in Yugoslavia as a military occupier would, they knew, threaten Greece and Italy and the whole eastern Mediterranean.

The Yugoslav Communists had not wanted to believe that Stalin or the Soviet Union would do anything to hurt them as long as they were on the same path toward a Communist world. They were sure they had not deviated from the path of Marxism-Leninism, and they did not think they were anti-Soviet or anti-Stalin. Writing several decades after the break, Djilas said: "Not a single party leader was anti-Soviet—not before the war, not during, not after." He says that "divergences began during the war. But our sense of intimate association with Moscow also stemmed from that period." On the other hand, he reports that "from our very first contact with Soviet officials and the Red Army in 1944 I had entertained doubts about Stalin and the Soviet system and wondered whether action can ever really coincide with principle." A less ideological view was expressed by Veljko Mićunović, like Djilas once a worshiper of Stalin, and twice Yugoslavia's ambassador to Moscow: "The Russians regard Eastern Europe as their own internal affair and, to judge from all the evidence, they will not need anybody's approval, and certainly not the Yugoslavs, for any solution they may decide on."

In 1948, Tito and his comrades felt the need to show that they were on the Marxist-Leninist path, and in the next year or so sought to prove it in various ways. One was a determined drive to accelerate the collectivization of agriculture. Another was to demonstrate that the Communist Party was in effect the dictatorship of the proletariat.

By the early 1950s, however, they realized that their actions would have no impact on Stalin. Consequently, they had to prove that it was Stalin who was the real deviationist. This meant admitting that they were wrong in copying the faulty Stalinist model. They needed to devise new or different political and economic forms and institutions in order to be true to Marxism-Leninism—or at least to Marxism. At its congress in 1952, the party changed its name to League of Yugoslav Communists to signify that, unlike the Soviet party, the LCY was not a command structure; it would seek to have its decisions adopted by persuasion. Many citizens viewed this as a distinction without a difference.

In the next decade or so, Yugoslav Communists engaged in considerable experimentation. Politically, some party leaders wanted to move faster, and this led to purges. Among the first to fall was Tito's close associate in the war years and in the conflict with the Cominform Milovan Djilas, who subsequently became Yugoslavia's leading dissident. Changes were introduced several times in legislative bodies, but with no real results. At the national level, the Council of Nationalities gave way to a new upper house, the Council of Producers, which was to represent various sectors of the economy. The lower house was the Federal Council.

After further changes, in 1971, there was a Federal Assembly with five chambers. In 1974, this was reconstituted to consist of the Federal Chamber and the Chamber of Nations and Nationalities (this last, to avoid using the term minorities). Members of both chambers were selected indirectly in accordance with what is called the delegate system. Changes were also made in the electoral law to permit more than one candidate to run for office.

In late 1971, Tito introduced the concept—referred to earlier—of a collective presidency, which originally was made up of twenty-three persons, but later was reduced to one from each of the six republics and the two autonomous provinces, who, with the head of the party, made a total of nine. Later still, the collective presidency was limited to eight. Each year these were to elect, in rotation, a president and a vice-president from among themselves. The implementation of this system was postponed until after Tito died because he had been elected president for life.

The rotation principle was also applied throughout much of society. Political factions had begun to form, not because of the artificial rotation scheme, but largely because of some political decentralization. One unforeseen result was that Tito found it necessary in 1971 to purge the Croatian Party leadership for its tolerance of nationalistic tendencies. To appear to be evenhanded, in 1972 he also purged the party leadership in Serbia. Less noticeable purges took place in parties in other republics.

By the mid-1970s, political liberalization was rearing its head in ideological circles. Eight university professors, mostly in the philosophy faculty at the University of Belgrade but also in Zagreb, known as the Praxis Group, were charged with espousing unorthodox Marxist views. To a large degree, it was their students who had been in the forefront of the student demonstrations of 1968, for which they got most of the blame. Their party unit was dissolved and a new one created, in which these professors did not seek membership; they continued their activities. The regime sought to get them ousted by a variety of techniques, but it was not until 1975 that they were dismissed, and that took an act of the Serbian parliament. They were allowed to pursue research work but could not teach. They were restored to their professorships only after Tito's death. In general, however, there was more freedom after 1948 in cultural and literary fields than previously.

The regime acted in other areas of concern. The 1974 constitution—following many throughout the preceding years—while providing for more autonomy for the republics and autonomous provinces, sought to reassert the primacy of the dictatorship of the proletariat. There was a general tightening of censorship, and in the 1970s several underground Cominformists were arrested.

In the economic sphere there were many changes. The years 1945 to 1953 can be characterized as administrative centralism. Agriculture was then decollectivized, and, in the decade that followed, a great deal of emphasis was placed on workers' councils and self-management. Yet the results were not as expected: Workers worked less and expected more money, many new bureaucracies were created, and there was a reduction in cost efficiency. The period after 1965 saw a series of endless reforms. There was to be less central direction; fiscal policy, money, and credit were to become more important as vehicles of production. A complex system of Basic Organizations of Associated Labor was inaugurated in the expectation that self-management would be improved thereby. But in the 1970s the economic situation began to deteriorate. Inflation was on the rise as well as prices. The trade deficit increased. Labor strikes, which were not supposed to occur in a workers' society, began to spread. Moreover, trials for misuse or embezzlement of enterprise funds received considerable publicity.

By the middle of the 1970s there was a strong feeling, especially among intellectuals, that Tito had lived too long, that no important changes could take place while he was in power. Because these sentiments could not be expressed publicly, considerable frustration grew among politically aware citizens.

On the international scene, after the death of Stalin in 1953, Tito was able in a few short years to reach a nominal rapprochement with Moscow. Although the country had been welcoming any aid, material or psychological, from the West, the leadership, not wanting to appear to be moving to the anti-Soviet camp, now said they would not accept aid if conditions were attached. They particularly wanted to demonstrate to Third World countries or movements that Yugoslavia would not be associated with either the pro-Soviet or the pro-Western camp. Tito wanted to show that he was bigger than life, that he could say no to the great powers. At the same time, he realized the weakness of Yugoslavia's international position if it were to be isolated. Out of this realization came the idea of nonalignment. Tito, India's Nehru, and other leaders of Third World countries therefore sponsored congresses of nonaligned nations for a number of years. In most of these, Tito was looked upon as the pioneer, and many representatives took their cues from him. Complicating the picture was Cuba's Fidel Castro, who managed to get himself into the nonaligned camp. Further, the delegates frequently voted for positions favorable to the Soviet Union and against the United States. A year or so before his death, Tito was embarrassed at the nonaligned congress that met in Havana by having his "nonaligned" position openly rejected. After his death in 1980, little homage was paid to him by leaders in the nonaligned camp.

9

Tito's Legacy and the End of Communism

Tito as dictator ruled over Yugoslavia for thirty-five years. What was his legacy? Briefly, he left the Yugoslavs a surfeit of problems, political, economic, and ethnic, and a political system ill-suited to deal with them.

Tito, the Croatian metal worker whose pseudonym in the Soviet-directed international communist movement was Walter, later changed to Tito, was the one Yugoslav Communist leader who survived the Soviet purges of the 1930s. World War II provided him with the opportunity to organize a guerrilla movement whose main purpose was the seizure of power. As noted in an earlier chapter, he was successful in portraying his movement as the major fighting force against the Axis in Yugoslavia, while at the same time hiding his intentions to introduce communism there. Consequently, the West gave him considerable aid, while abandoning the leader of the first resistance in Yugoslavia, Draža Mihailović, whom Tito executed at the end of the war.

His brutal communist dictatorship was modified to some degree after his break with Stalin in 1948, but never to the extent of endangering Communist Party rule. Similarly, in the economic sphere, the people were told to manage their own affairs (the highly touted self-management), but the party told them how to do it.

The nature of Tito's dictatorship must be kept in mind in any effort to assess the performance of the political system that he bequeathed to his heirs.

Contrary to the belief held in some circles, Tito's revolution was not mild or benevolent. In its initial years it was fully as brutal as the Soviet or any other Communist revolution. As indicated in an earlier chapter, Tito's Partisans never rejected any tactic that might lead them to power. When they came to power, they were determined to eliminate anything and everything that stood in the way of their imposing dictatorial rule in order to create a new society. There was killing to get power and killing to keep power. Individuals and organizations (churches for example) were shown no mercy. The one-track approach led not only to painful and costly dislocations, but also to distrust and division among groups, families, and communities. Pitting neighbor against neighbor or nationality against nationality was acceptable if it served the Party's interests. Past habits, such as saving for one's children, or elementary decency rooted in ethical mores of the common people, were undermined or brought into question. Indeed, Tito's demolition of the South Slav peoples' moral fiber may turn out to be his most enduring legacy.

It can be said that every revolution has had its excesses. Most revolutions, after relatively brief periods, consolidated their gains and the repressions of the early years moderated or disappeared. This is not true of Communist revolutions. Communist regimes never feel secure and therefore rarely relax their restrictions. The difference can perhaps be explained by the fact that past revolutions were usually limited in their objectives, whereas Communist ones pursue a goal of completely remaking society. Hence the term totalitarian to describe them.

After 35 years of Tito's rule, with its changes and controlled liberalization, plus the general resignation of the populace, few observers expected the downfall of the system. Yet the regime failed to generate confidence in its ability to cope with the many problems facing the nation.

Tito's heirs, gathered in the 23-member Presidium of the League of Yugoslav Communists and the collective presidency of eight, swore that they would not deviate from "Tito's Road," but this proved to be more difficult than they thought. Instead, the LCY experienced a steady decline in its influence and authority. While Tito lived, he was the supreme arbiter when disagreements arose in high party circles, but no such referee was available once he left the scene. His heirs were saddled with a collective of equals.

The LCY quickly became a coalition of increasingly powerful republic and provincial politicians. There was no longer a Yugoslav Communist Party, but eight different parties. The principle of "democratic centralism" was retained to suggest party unity, but it was watered down, and disagreements erupted as to its meaning. In practice it became a process of endless discussion that bound no one. Some called it a "voluntary obligation."

In theory, the LCY was still dominant, but in practice its function became that of seeking agreement or consensus, while at the same time hiding from the public the widening gulfs between the eight party camps. At the same time, LCY leaders purported to be struggling for ideological unity. A resolution of the central committee in March 1985, for example, declared that its task should be to "establish and formulate a program for analyzing and resolving the most important issues," emphasizing democratic centralism, which must not degenerate into "sterile repetitions" of the same resolutions. At the same meeting, the central committee put forth the following mouthful: "Political Platform of Action by the LCY in Developing Socialist Self-Management, Brotherhood and Unity, and Togetherness in Kosovo."

In 1987, there was a special meeting of the central committee on ideological questions. At that and at other central committee conclaves, there were ringing calls for eliminating conflicts and restoring unity. The response was skepticism. LCY congresses echoed the same generalizations and produced mountains of resolutions: they agreed on the problems but not on solutions. In 1988, there was a special party conference, whose lack of concrete action merely strengthened the realization that the party had lost control over the national-policy agenda. Some observers noted that the congresses were simply arenas for the clashing leaderships of the individual republics.

By 1989, republic and province parties began to quarrel publicly, and people began to view the conflicts as expressions of local interests. In addition, the party began losing members, especially among the young. Party leaders were criticized for ineptitude, corruption, and abuse of power. There were charges of opportunism and challenges by specific movements related to ecology, feminism, pacifism, and other issues.

Criticism of party policies led to ever wider attacks. In 1986, for example, the prestigious Serbian Academy of Sciences and Arts began drafting a memorandum entitled "The Crisis of the Yugoslav Economy and Society." It was, however, seized in draft form and severely censured for being motivated by the spirit of "Serbian hegemonism." The most sensitive part of the memorandum dealt with Serbian grievances against the Yugoslav Communist Party, against Stalin and the Comintern, and against non-Serbian party leaders, including Tito and Vladimir Bakarić, Croats, and Edward Kardelj, a Slovene. The academy defied the LCY, gave full support to its own leaders, and denied later charges that it had incited the growth of Serbian nationalism.

Divisions in the party had their counterparts in the government, and were particularly troublesome in view of the requirement of near unanimity to reach decisions on substantive issues. The 1974 constitution required that certain laws must receive a two-thirds vote in the Chamber of Republics and Provinces. In case of emergency, action could be taken by a curious body made up of the LCY secretary, the president of the parliament, the prime minister, the minister of foreign affairs, the minister of defense, the minister of the interior, and the state presidency of eight. It is clear that this process was more than cumbersome; it was conducive to paralysis.

Simultaneous with these developments was the worsening economic situation. For a time in the late 1960s and early 1970s, Yugoslavia experienced some prosperity, due in large part to the infusion of credits by the West, and the toleration by the regime of people's yearning for a consumer society. By the time Tito died, however, the chickens were coming home to roost and the standard of living had begun a decline that continued thereafter.

The problem was that the Yugoslav economy never operated at a profit. It always needed a transfer of resources to keep it going—from U.S. grants in the early 1950s to subsequent heavy borrowings, aided by remittances sent home monthly by almost a million Yugoslav workers in the West. Fundamentally, there was an insurmountable discrepancy between "self-management" and "social ownership of the means of production," on the one hand, and the drying up of foreign credits in the 1980s, on the other.

In the decade after Tito's death, several different economic packages were tried, but with singular lack of success. The problems of inflation and unemployment were exacerbated by scandals and strikes. Shortages were followed by rationing of some commodities, then price and wage controls. Some enterprises were weeks and even months behind in the payment of wages to their workers. The currency was devalued several times. In the late 1980s, there was an effort to make the Yugoslav currency convertible to the German mark, but in this the republics were far from cooperative. Moreover, the banks were drained of their hard currency reserves in their effort to meet the needs of convertibility. Foreign debt reached about twenty billion dollars and debt servicing became the albatross of the Yugoslav economy.

Strikes increased in frequency, although in a Communist system, in which everything was held in common, the workers would be striking against themselves. Initially, they were called unauthorized work stoppages. Subsequently, they were recognized as legitimate, first in Slovenia and Croatia in 1988.

Even in Tito's day, the central authorities had used some market mechanisms, and the system was sometimes referred to as "market socialism." The 1974 constitution, however, greatly limited the central government's role in the sphere of economic policy. The resulting decentralization, which had begun in the 1960s, often led to ruinous competition among the republics, as each felt the need to produce items that were produced in neighboring republics, even at uneconomical cost. There were impediments to the free flow of goods. When a train came to the frontier of a republic, for example, it was necessary to uncouple the engine and replace it with one from the next republic and so on to the end of the line. Other nationalistic restraints were common.

Movement of capital across republic borders was always looked upon by some as a net loss of resources. In addition, the movement of goods was impeded by restrictions that favored one republic's products irrespective of price or quality. Party leaders in each republic had vested interests in keeping certain enterprises operating, even if they were losing money. In popular parlance, such enterprises came to be known as "political factories." These things could take place because there was no unified Yugoslav market.

New plans for economic stability were usually identified with incoming prime ministers, but prime ministers came and went, and economic dislocations grew apace. After 1980 a significant shift of trade toward the Soviet bloc was tried, with no marked improvement; it merely reflected the inability of the Yugoslav economy to compete where it really mattered—in hard-currency Western markets.

During these years, the government continued to declare itself in favor of general liberalization, but impeded or banned books, newspapers, and magazines, and criticized dissident writers. At the same time, paradox upon paradox, the government gave permission to the Serbian Orthodox Church to complete Saint Sava Cathedral. Its construction had begun before World War II, but Tito would never allow it to be completed.

Tensions between republics grew. The Slovenes and the Croats insisted that they were contributing much more to the central treasury than they were getting back, that for their hard work they were being exploited by the lazy and inefficient republics. They ignored the fact that they had benefited from local investment during the First, as well as the Second, Yugoslavia at the expense of other regions. Croatia, with its large hard-currency earnings from tourists, took little account of the fact that it fed its tourists on large quantities of foodstuffs from other republics, mainly Serbia. The Croats and Slovenes also benefited by being able to buy inexpensive electricity and raw materials from the southeastern part of the country—Serbia and Macedonia—and to sell their value-added industrial products back to those regions without internal customs barriers. This was reminiscent of the north-south situation in the United States at the time of the Civil War.

The West has in recent years heard the arguments of Slovenes and Croats that they have been economically exploited, but their charges of exploitation would seem to be far from convincing in view of the fact that in 1990 the average per-capita income in Slovenia was $12,618; in Croatia, $7,179; in Serbia, $4,870. Serbians averaged a per-capita income that was only one-third that of Slovenians.

Marx had taught that capitalism was responsible for national or ethnic conflicts. Accordingly, with the inauguration of Communism, Tito and his comrades had declared the nationality problem solved, and they provided criminal penalties for those who disagreed. The most dramatic evidence that the nationality problem was far from solved was Kosovo, the autonomous province created out of the republic of Serbia.

It had been the center of the Serbian kingdom of the Middle Ages, when Serbia was the strongest empire in the Balkans for more than a hundred years. It is there that the most cherished Serbian Christian monuments, its monasteries and churches, are located. It was there that the Serbian army suffered defeat in 1389 at the hands of the Ottoman Turks, and the Serbian monarch, Prince Lazar, lost his life. In the nearly five hundred years of Turkish rule that followed, the Serbs never stopped dreaming of a resurrected Serbian state, a dream that was realized in the nineteenth century. In Montenegrin epics—the last true folk epics that thrived in Europe—the deaths of Lazar and his heroic leader, Miloš Obilić, were sung to the accompaniment of the Serb national instrument, the gusle, from 1389 to 1992, and no doubt will persist for centuries more.

The holy ground of Kosovo was not regained until the Balkan Wars of 1912. During the intervening centuries, many Serbs fled Kosovo to escape enslavement, while the Turks sent in more and more Islamicized Albanians, who engaged in massive persecutions of Serbs, which today would be called genocide, and which were reported in detail by consuls of the European powers. These persecutions were particularly odious in the nineteenth century. The net result was a sharp increase in the percentage of Albanians in Kosovo. The high birthrate among Muslim Albanians also contributed to that increase.

After Mussolini's takeover of Albania in 1939, and the

defeat of Yugoslavia in 1941, Italy enabled the Albanians to create a Great Albania, which incorporated Kosovo. Following the defeat of Italy in 1943, the Kosovo Albanians sent a pro-Nazi regiment to fight on the side of the Germans. At the same time, Tito, as leader of the Partisan resistance, in a desperate effort to get help wherever he could find it, offered advantages to the Kosovo Albanians. He got precious little help from them, but he allowed at least a hundred thousand Albanians to move to Kosovo, making the balance between Serbs and Albanians nearly even.

Tito also had promised the Kosovo Albanians that they could be annexed to Albania, and although he reneged on that promise, he did allow them to form an autonomous province. The Kosovo Albanians lost no time in seeking to make of Kosovo an ethnically pure area. To that end, they engaged in continual and brutal persecution of the Serbs: They raped and pillaged; desecrated Serbian religious institutions, including cemeteries; set Serbian barns and haystacks on fire; cut timber on Serbian lands, and constructed buildings on Serbian property. The purpose, of course, was to force Serbs to flee, and they did so in large numbers.

Local Communist authorities looked the other way or, worse, actually conspired with the Kosovo Albanians. Protests of Serbs to local and federal authorities were to no avail. Although an autonomous province of Serbia, Kosovo was ruled as if it were a sovereign state. One indication of this was the importation from Albania of more than two hundred professors and countless textbooks for the University of Priština, originally a part of Belgrade University and therefore funded by Serbia.

A year after Tito's death, the Kosovo Albanians launched

large-scale demonstrations, demanding the status of a republic and even the right to be annexed to Albania. Not
until then had informed people in Belgrade dared mention
the suffering of the Serbs in Kosovo. A notable exception
was the great Serbian novelist Dobrica Ćosić. Although the
demonstrations were put down by force, the situation for
the Serbs continued to deteriorate.

In April 1987, Slobodan Milošević, who a few months
earlier had become head of the Communist Party of Serbia,
went to Kosovo to listen to the Serbs. He got an earful. The
meeting lasted thirteen hours; seventy-eight persons spoke.
The newspapers in Belgrade thereafter began openly to speak
of genocide, and to express amazement that in the past six
years there had not been a single political resignation in Kosovo—nor at the top of the Yugoslav government—as an
indication of the acceptance of responsibility for this.

In June, the LCY concluded that the "most difficult part
of the problem of Kosovo and the whole of Yugoslav society
is to be found in that the policy of the LCY is not being
implemented." It did not say who was failing to implement
party policy. A month earlier, at an "ideological" meeting
of the central committee of the LCY, a member asserted: "If
we cannot quickly overcome genocide," then we should call
for "free elections, with multiple candidates, so that men
can come to the top who can bring an end to the genocide."

Milošević vowed that he would not permit anyone to
abuse Serbs the way they had been mistreated in Kosovo.
This act won him a great number of supporters among Serbs;
even the most anti-Communist among them looked upon
him with at least grudging admiration. As Serbs studied the
nationality policy of the Communists, they were horrified to

learn that the Serbs of Kosovo, whose percentage of the population had shrunk to about twelve, were not even entitled to the minority-rights guarantees of the Yugoslav constitution. The explanation is simple. The constitution recognizes nations and nationalities. Nations are the major groups that have their own republics (for example, Serbia, Croatia, Slovenia); they are not minorities. Nationalities are groups (for example, Albanians, Hungarians, Italians) that do not have their own republics, which makes them minorities. By constitutional definition, therefore, Serbs who lived outside Serbia were technically not minorities. Consequently, the Serbs of Kosovo could not invoke the minority-rights guarantees of the national constitution. The same was true of the Serbian minority in Croatia.

The problem of Kosovo deserves attention because the events there led many Serbs, who had been the strongest supporters of the Yugoslav state, to question that long-held commitment. Adding to this change in attitude was criticism by other republics, notably Slovenia and Croatia, of Serbian actions in Kosovo to protect the Serbian minority. The Serbs felt that the leaders of their brother republics, who had helped the Albanian authorities in Kosovo, had sold out to separatism. Understandably, they feared the destruction of historical Serbia.

Milošević's dramatizing of the Kosovo question opened the way for Serbs to ventilate many other grievances. They began pointing out that Serbs had sacrificed far more than others for the common state, yet they were being blamed for all the failures of the First and Second Yugoslavias. Moreover, they remained distressed by the fact that a third of them were forced to live outside the borders of Serbia and

Montenegro, and for years they had been prevented from restoring the reputation of their guerrilla leader Draža Mihailović.

While top leaders of the LCY sought to present an optimistic outlook to the public, they had secret doubts. Other citizens, even some LCY members, became more vocal. The process of de-Titoization, which had been slow in starting, gained momentum. Many found him to be the real culprit for the country's woes. Articles began appearing that described his opulent life-style, his luxurious homes in various parts of the country, and his exorbitant spending of the people's money to impress foreign guests.

Tito's heirs, in what became their final act in search of his Communist road, in January 1990 convened an extraordinary congress of the LCY. It failed to resolve differences. There was agreement that the party should give up its monopoly of power and that a multiparty system should be adopted. Before this and other resolutions could be formalized, however, the Slovene delegates walked out, a move supported by the Croats. The congress was adjourned indefinitely, and, for all practical purposes, the once-proud Yugoslav Communist Party had breathed its last.

Following ominous signs during the previous years, there occurred in 1989 the extraordinary collapse of Communist regimes in Eastern Europe. The impact in Yugoslavia was dramatic. Onetime Communist leaders began to realize that Communism had lost the East-West Cold War and had little appeal to ordinary citizens, even those fearful of change. Milošević had recognized the appeal of nationalism when he stood up for the rights of Serbs in Kosovo. The lesson was not lost on the Communists in Slovenia and Croatia, who began casting their demands for change in nationalist terms.

Slovenia and Croatia held elections in the early part of 1990; Bosnia-Herzegovina, Macedonia, and Montenegro, during the year; Serbia at the end of the year. In the first two republics, non-Communists won, largely on the basis of nationalist slogans, even though the leaders were recycled Communists. The man who became president of Croatia, Franjo Tudjman, had been a Communist Partisan general in Tito's army. The Slovenian president, Milan Kućan, also had a long and notable Communist past. In Serbia and Montenegro, the Communists were victorious, but they quickly replaced the word "Communist" with "Socialist." In Bosnia-Herzegovina, the elections resulted in a three-way split, closely proportionate to the population division of Muslim, Serb, and Croat. In Macedonia, the recycled Communists established an uneasy coalition with the ultranationalist V.M.R.O. (pre-Communist Macedonian Revolutionary Organization), which captured most of the seats in the assembly, with the V.M.R.O. becoming the largest single party. In Kosovo, the large Albanian majority boycotted the elections, perhaps fearing that if they participated, they would be recognizing that they were a part of Serbia.

The Communists in power in Montenegro and the autonomous province Vojvodina (part of Serbia) were subjected to massive demonstrations and forced to resign. Most observers believed that Serbia's Milošević initiated or authorized these moves in some way, particularly when the ousted leaders were replaced by men loyal to him. In response to strikes and other disturbances by Kosovo Albanians, Milošević succeeded in getting the Serbian constitution amended so that the vital functions of the Kosovo government (police, defense, courts, internal order) came under direct Serbian control. For the first time since the inauguration of the 1974 constitution, Serbia could be said to have a decisive voice in the affairs of its two autonomous provinces, regaining part of a nation that was whole under the Kingdom of the Serbs, Croats, and Slovenes.

It should be remembered that the Kosovo Albanians had been able to participate fully and equally in the political affairs of the Second Yugoslavia. They had their share of diplomatic posts abroad, as well as posts in the legislative, executive, and judicial branches of the national government. Moreover, they had a virtual veto in the policy-making of the government of Serbia. Ironically, they often had more of a voice in the Serbian republic's affairs than Serbia had in internal Kosovo matters.

Slovenia and Croatia were quick to charge that Serbia's moves to exercise control over its autonomous provinces were but further proof that Serbia was determined to dominate the state. Once again the cry of "Great Serbia" was heard, and it became a propaganda weapon in the hands of Croats and Slovenes who spoke to Western journalists. Both seemingly forgot the policy, under Tito and after, that established the "national key" for filling government positions: The republics and provinces were to be represented roughly in proportion to their populations. By and large, the national government had adhered to this principle, under which the Serbs could in fact charge discrimination.

Key economic posts, for example, were held mostly by Slovenes. And whereas Serbs may have held their share of ambassadorships, a Serb was rarely sent to head a Yugoslav embassy in one of the major world capitals. In the post-Tito decade, the prime ministership was held only once by a Serb, and he was from Bosnia-Herzegovina. In the critical year 1991, Serbs did not hold the prime ministership, the foreign affairs ministry, or the chairmanship of the collective presidency; the minister of defense was General Veljko Kadijević, the son of a mixed Serb-Croat marriage.

These facts were brushed aside in the war of words as Slovenia and Croatia, in 1990, began demanding a reorganization of Yugoslavia that would result in a loose association of sovereign states. Some Slovenes and Croats spoke of a confederation. These demands were increased in early 1991, with declarations that if their requests were not met, Slovenia and Croatia would secede. They pointed out that the Yugoslav constitution gave them the right to do so. This right of secession is, however, complicated. In the preamble to the constitution there is mention of the right of self-determination, including the right of secession, but nowhere in the text does it explain by what procedure this right may be exercised.

Yugoslav constitutional experts point out that certain provisions make quite clear the spirit of the constitution. For example, Article 5 says that the frontiers of Yugoslavia may not be altered without the consent of all republics and autonomous provinces, and it stipulates that boundaries between republics may be altered only by mutual consent. In addition, Article 240 says that the Yugoslav armed forces shall protect the independence, sovereignty, and territorial integrity of Yugoslavia.

Prior to the acts of secession, the Slovenes and Croats

proposed a new constitution. Under it, the national government would have consultative powers, making it seem even weaker than the national government of the United States under the Articles of Confederation (1781–89). Under such a system, Serb interests anywhere outside Serbia could not be protected by the national government or the government of Serbia. With more than thirty percent of the Serbs living outside Serbia and Montenegro, it seemed highly unlikely that Serbia could ever agree to such a constitution.

Under such a constitution, or in the event of secession, there would be a need for redrawing boundaries. This would open up many new problems, particularly with respect to Bosnia-Herzegovina and with the question of autonomy for minorities elsewhere.

In the political stalemate of midsummer, Slovenia held a referendum, which indicated the will of the people to secede. A similar referendum was conducted in Croatia, with a similar result. Subsequently, Slovenia declared that it was seceding from Yugoslavia. This was followed by a similar announcement by Croatia. The Yugoslav Constitutional Court, however, nullified the key provisions of the Croat and Slovene declarations of secession. The central Yugoslav government avowed that those actions were illegal and sent troops to control border crossings in Slovenia, with no small amount of bloodshed as a result. After a cease-fire agreement was signed, with the assistance of the Western European powers, federal troops were withdrawn.

More bloodshed occurred in Croatia, between Croatian militia and armed groups of Serbs representing the 600,000 to 800,000 minority Serbs living in Croatia. These Serbs were the descendants of the survivors of the massacre of 1941 and, going further back, descendants of those who protected Austria-Hungary's citizens (including Croats) on the military border, the Krajina, from the Ottoman Turks. Soon the Yugoslav army got involved, and was, by many Western press centers, accused of being an instrument of Serbia. The army's officer corps was predominantly Serbian, yet in 1990 the high command was 38 percent Croat, 33 percent Serb, 8.3 percent Slovene, 8.3 percent Macedonian, and 4.1 percent Muslim—hardly proportionate in a country in which almost half the population was Serb.

It should be noted that, although the army may have at times followed Serbia's leadership, it seemed for the most part to be acting on its own, or as if it was fulfilling the will of late dictator Tito, who had often declared that in times of trouble the army could be counted upon to save Yugoslavia.

Serbs in and out of Croatia had not lost sight of the fact that when Croatia was an Axis satellite during World War II, the regime was responsible for the massacre of some 500,000 to 700,000 Serbs. Now, they were frightened by the fact that the new Croatian regime was copying some of the symbols of the hated wartime Ustashi. Serbs in Croatia were also being systematically dismissed from their jobs, and were facing other kinds of discrimination.

These concerns led their leaders to hold a plebiscite in their areas of Croatia: Krajina, on Croatia's border with Bosnia-Herzegovina, and Slavonia and Western Srem, in which at least a third and in some areas half of the population was Serb. The result indicated that should Croatia remain part of Yugoslavia, the resident Serbs would be satisfied with cultural autonomy; if Croatia was to secede, the Serbs would be satisfied with nothing less than annexation to Serbia. Despite this warning, Croatia's leaders declared their secession.

This development was a morally depressing blow to Serbs throughout Yugoslavia and to Serbia. Although the formation of Yugoslavia in 1918 was a voluntary act, the Serbs had taken the lead in its creation and in many ways had sacrificed the most for the common state. Even when Yugoslavia became Communist, which adversely affected the Serbs and enlarged Croatia's borders, they remained loyal to the concept of a united Yugoslav state. It is not surprising, therefore, that the Serbs tasted gall, a bitterness rooted in countless betrayals. That bitterness seemed all the more cruel when the Serbs reflected on the fact that during the past half century, and for the first time in their one-thousand-year history, their fate had been determined by a person who spoke their language and counted on their military support though he (Tito) was of a different faith and a different nationality.

Tragically for Yugoslavia, and although Communism was clearly on the decline in the last years of Tito's regime, the opponents of Communism could not develop a united front. Dissidents probably spoke for an overwhelming majority in all the republics, but they failed to reach agreement on what was to be done in their country of twenty-four million people, a country that could, instead, play a significant role in Eastern Europe. Those in Serbia were unable to evoke a cooperative response from Croats and Slovenes.

Dissidence had first appeared in Serbia when Tito's one-time close comrade Milovan Djilas wrote the book that was published abroad in 1954 as *The New Class*. For this, along with later publications, he spent years in prison. By the time of his last release, in 1967, dissent was less dangerous. Activist students at Belgrade University, reformist Marxist professors, and other opponents of the Communist system tried to find common ground with Croatian and Slovenian intellectuals, but they were on the whole unsuccessful. It is worth remembering that in the 1930s, important Serb opponents of the Yugoslav regime had cooperated with the Croats who led the United Opposition. Yet, ironically, in the 1970s and 1980s Serb dissident pleas for a united front fell on deaf ears in Slovenia and Croatia. Both Slovenian and Croatian Communists had turned to nationalism, having seen that such an appeal had worked for Serbia's Milošević when he defended the Serbs of Kosovo from persecution by the Kosovo Albanians.

And so, by the time the Communist system was in its death throes, narrow ethnic interests were triumphant.

10

The Dream Not Realized

What can be learned from the tragedy of Yugoslavia? Perhaps the most obvious answer is: Those who contributed to the formation of Yugoslavia were motivated by expectations that were unrealistic. Yugoslavia may have been a dream that could not be realized, or, at best, could be only imperfectly attained. Contrary to the views of some historians, however, it was not an idle dream; nor was Yugoslavia an unnatural or artificial creation. It came about as the result of the reasoned thought and forceful action of a number of South Slav writers and politicians during the latter half of the nineteenth century and the first two decades of the twentieth.

The formation of the state came about hurriedly, without concrete preparation. Romanticized visions and dreams smothered practical considerations. Those most responsible tended to ignore, or at least to underplay, the political, cultural, and religious differences that had always existed among the Serbs, Croats, Slovenes, and minorities in various parts of the country. In addition, they did not sufficiently take into account the fact that the Yugoslav idea was mainly conceived and propagated by middle-class intellectuals, who did not understand the peasantry, which represented more than eighty percent of the population. In the idealism of the moment, too many things were taken for granted, or were not given serious consideration.

This is understandable, however. The formal steps toward unification were taken when military pressures did not allow for deliberate discussion, public debate, and mature reflection. This was true of the Corfu Declaration and the Act of Union, as evidenced by the differences and disagreements that were not slow in appearing and that grew into problems that played no small part in the political life of the First Yugoslavia.

Leading South Slavs were not unaware of these circumstances. For one, Serbia's prime minister Nikola Pašić, when in late 1918 he was simultaneously accused of being lukewarm to the creation of a Yugoslav state and of seeking a predominant place for Serbia in that state, responded by saying: "I solemnly declare that Serbia views it its national duty to liberate Serbs, Croats, and Slovenes. When they become free, their right to self-determination will be recognized, that is, the right to declare freely that they wish to unite with Serbia on the basis of the Corfu Declaration, or that they wish to create small states as in the distant past." This cautionary note was interpreted by his critics to mean that he really wanted to form a Great Serbia instead of a Yugoslav state.

Moreover, at the very time that representatives from the National Council in Zagreb were preparing to go to Belgrade to consummate the Act of Union, some Croatians, notably Stjepan Radić, voiced warnings, though these were ignored. The National Council represented not only the

Croats, but other South Slavs in the rapidly dissolving Austro-Hungarian Empire, large areas of which were being overrun by Italian troops seeking to claim areas promised under the 1915 Treaty of London. It therefore did not have the luxury of debating self-determination or any other complicated political question.

When the Act of Union was signed on December 1, 1918, the First Yugoslavia vested royal authority in the hands of Prince Regent Alexander. Governmental authority was in the hands of the provisional parliament and the cabinet. These authorities faced great handicaps in trying to get a constitution for the new kingdom. They had to deal with the myriad problems of a country badly ravaged by four years of war and in which poverty and disease were endemic. Moreover, they were seeking to do this through democratic institutions, with which many were not familiar. The many difficult situations and endless recriminations of those days were but a foretaste of what was to come.

Croatia proved to be an early hurdle facing the new leaders. After the Croatian Peasant Party won the 1920 election for delegates to the constituent assembly, its leader, Radić, declared that he did not recognize the Act of Union and would work instead for an independent Croatian state. He and his delegates refused to participate in the making of a constitution for Yugoslavia. Under the Habsburg monarchy, the Croats had practiced noncooperation and obstructionism, too, tactics they easily transferred to the new state.

The Serbs were baffled. They had experienced bitter political battles, but none that had brought the existence of the state into question. The attitude of the Croats, especially Radić, reminded many Serbs of the fact that, willingly or unwillingly, the Croats had fought in the Austro-Hungarian army and served in Vienna's police forces and bureaucracy. They wondered how to explain the Croat action, now that they were free of Vienna's control and had expressed their desire to unite with Serbs and Slovenes to form a common state.

Influential Serbs suspected that the Croats could not easily rid themselves of the lingering influence of political and religious fanatics, but expected this condition to be only temporary. Moreover, the Serbs knew that the purpose of a political party was to struggle to gain power, and were convinced that the Croatian Peasant Party's aim was simply that. Sooner or later, that party would take its place in the political arena instead of being an antistate instrument. So they hoped. The Croats' acceptance of the political system in 1925 and their entrance into the cabinet seemed to demonstrate the correctness of this evaluation. Yet within two years the Croats resorted to unbridled opposition and obstructionist tactics in parliament, and Serbian hopes and expectations were shattered.

Today it is easy to conclude that the Serbs were naive, but it is well to remember that at the end of World War I the view was widely shared throughout the Western world that acceptance of democracy and democratic constitutions was the solution for political problems. This was Woodrow Wilson's credo. That this confidence was misplaced is another matter.

Another defect in the First Yugoslavia was cabinet instability. In the period of free elections prior to 1929, the average life of a cabinet was five months. Proportional representation was in large part responsible for the inability of any party to win a majority of seats in parliament. As in other democracies under similar circumstances, the prime ministers spent an inordinate amount of their time and energy currying favor with minority groups in order to maintain an uneasy coalition and stay in power.

Despite the weakness of the cabinets, the opposition was never able to cooperate to oust them. Except for a few months in 1924, Radicals, in various coalitions, held power from 1921 to 1929, due in large part to the abstention of the Croatian deputies. Croatian and Serbian opposition forces held many discussions, but they could never agree on a common plan. The closest they came to agreement was when Maček headed the United Opposition electoral lists in 1935 and 1938. That they failed to agree on a common political platform may have been due to the Croats' fear that they might then endow the system with legitimacy by their participation.

Throughout, it is fair to say, the Croats preferred to avoid broad united associations in favor of organizations in which they could stress their national program. While the Serbian oppositionists were far from united, their opposition was essentially political, whereas that of the Croats was national. This explains why in 1939, as Europe made ready for World War II, Maček could conclude an agreement with Belgrade and why that act caused consternation among the Serbian members of the United Opposition.

On the other hand, even if a party could have achieved a majority in parliament, would that have been sufficient in the complex circumstances of the day to make a constitution and to govern? No one can know. The Serbs thought the Croats were afraid they might settle for less than the Serbs might concede. A distrust of all cabinets was fanned not only by Croats, but also by the Serbian opposition. In addition, Serbian ministers and their defenders wittingly or unwittingly sometimes conveyed the impression of superior knowledge, and since the Croats suffered from political inferiority, fed by their numerical weakness, this led them to suspect even the most innocent of Serbian moves.

Another weakness of the First Yugoslavia was lack of sufficient communication. Serbian leaders, both government and opposition, did travel to various Croatian regions in an avowed effort to improve communication on common problems, and King Alexander made a point of visiting Croatia and other parts of the country. Croatian leaders, on the other hand, traveled little to Serbia and far less to Montenegro and Macedonia. When they did go to Belgrade, it was to see the king, not to exchange views with Serb leaders, writers, or professionals. Serbian oppositionists failed to get Croatian politicians to visit their districts. It is sad to think that Maček made only one trip to Serbia outside Belgrade; that was to Kragujevac.

Through all political events of the First Yugoslavia ran a two-stranded thread: Serbian politicians did not have a determined political line directing their relations with Croats; the Croats had a constant Croatian line. The Croats reduced their five or six political parties of relatively equal strength to one, a Croatian national party. The Serbs moved from two major parties to about ten, to say nothing of the factions that developed within some. The Slovenes were divided, but in the main believed that Slovene interests could best be protected by participating in the governing of the state. The Yugoslav Muslim Organization also believed that retaining its special position in Bosnia-Herzegovina could best be assured by having its representatives in the cabinet.

The First Yugoslavia cannot be judged without considering the mind of King Alexander. The available evidence suggests that he was a sincere Yugoslav and did not favor one ethnic group over another. Because he believed it was necessary for all citizens of the country to think of themselves as Yugoslavs, and only secondarily as Serbs, Croats, or Slovenes, he changed the name of the country from the Kingdom of the Serbs, Croats, and Slovenes to the Kingdom of Yugoslavia. He thought that he could advance the common cause by temporarily assuming personal power, a step that did not win over the Croats and brought additional division among the Serbs.

Internationally, Alexander worked to strengthen Yugoslavia and to that end entered into alliances designed to neutralize neighboring nations that hoped the Yugoslav state would fail.

While he and the various cabinets were steering Yugoslavia through murky domestic and international waters, opposition forces often seemed more interested in bringing the governmental machinery to a screeching halt than in proposing realistic solutions.

What would have followed the sporazum of August 1939 if World War II had not intervened cannot be known. It seems evident that this agreement between Serb and Croat leaders to try to solve the Croatian question came too late, and at a time when the international situation suited Yugoslavia least. Many Yugoslavs believed, or at least hoped, however, that it was a concrete step toward a viable political system, but the war foreclosed further steps.

A final point about the First Yugoslavia concerns the role of alleged Serbian hegemony in the failure of the state to achieve a viable political system. My investigations have led me to conclude that those who have used this argument have not produced believable evidence. The Serbs may be blamed for much, but they were not guilty of economic exploitation of other groups, or of using their political position for their own material enrichment or the subjugation of other peoples. If anything, they exploited themselves for the benefit of the nation as a whole. True, there was corruption, mainly conflicts of interest or the use of privileged information, but this benefited only a small circle of persons, and they were not exclusively Serbs.

It is also true that during most of the two decades Serbs held the key positions in the central government. To a large degree this was forced on them, first by the refusal of the elected Croat representatives to participate in governing the country, and, still later, the Croats' penchant for engaging in obstructionist tactics.

If it can be said that the result was Serbian hegemony, it was the least planned, least conscious, and least profitable kind of hegemony. One of the sharpest critics of Serbian-led cabinets, and a close collaborator of the Croats, Dragoljub Jovanović, reflecting on events decades later, concluded: "Serbia is neither Great Serbian nor hegemonistic . . . it is not against Croats . . . neither revengeful nor capricious."

Unfortunately, many journalists as well as scholars have used the term "Serbian hegemony" or "Greater Serbia" as a shortcut to explain or categorize certain political events. Neither term was ever used in this way during Tito's regime; it was resurrected in the post-Tito period to explain interethnic conflicts, with particular reference to Serbian leader Slobodan Milošević. But as in the past, the terms do not shed much light on the causes and effects of historical occurrences.

The Second Yugoslavia, or Tito's Yugoslavia, grew out of a bitter and bloody civil war that had pitted citizens against citizens, and produced doubts, discord, distrust, hatred, and fear. Its Communist rulers promised a new beginning, a "people's democracy," a progressive society motivated by "brotherhood and unity."

The introduction of this dictatorship of the proletariat was hardly auspicious. With its secret police and tight control over all phases of human endeavor, the government determined to prevent even the most innocent dissent.

Central here is Tito's treatment of the largest ethnic group, the Serbs. Quite aside from his decision to scatter the Serbs, so that more than thirty percent lived outside the boundaries of Serbia and Montenegro, Tito, during his tenure as dictator, was never able to bring himself to recognize Serbia's merits in any field, except indirectly. Late in his life, for example, when he dedicated the Belgrade-Bar railroad, long a dream of the Serbs of Serbia and Montenegro, he could not utter the word "Serb" or "Serbia." He referred to the railway as the dream of "our" people, and as an extraordinary achievement of "our" people, but never the Serbian people.

It is not irrelevant to note here that in the civil war of 1991–92 between Serbs and Croats there has been no mention in the Western press that Serbia is still for the most part landlocked. The road north from Bar, on the southwest coast of Montenegro, passes through mountainous country but is distant from Novi Sad, in Vojvodina, and from Belgrade.

Serbs in Krajina fought, unsuccessfully, over Zadar, on the Adriatic, in part because it was deemed to be their area, but also for the economic benefit of gaining access to a port.

Fundamentally, the Communists tried to remake society in too much of a hurry, even had the populace been receptive. One of Tito's closest collaborators noted in the first days of the regime that in ten years Yugoslavia would overtake America! The effort to industrialize rapidly was unrealistic, given the lack of sufficient raw materials, adequate communications, and a skilled labor force. Yugoslavia was overwhelmingly a peasant country, and encouraging the peasants to move to new industrial sites created housing and attendant problems, as well as an agricultural labor shortage.

The arbitrary efforts to collectivize agriculture, penalizing unwilling peasants, robbed the country of a great deal of what had been the basic source of capital formation. Moreover, these efforts led to food shortages, in a country that in the prewar period had been an exporter of agricultural produce. Fortunately, the leaders early saw their error, prodded to some extent by Western powers eager to see Tito withstand the efforts of the Soviets to overthrow him. Thus, in 1953, they began the decollectivization of agriculture.

On the positive side, it can be said that the Communist regime built much-needed roads and railroads, improved river navigation, and constructed hydroelectric dams as a necessary energy source. Much of this would not have been possible without a great deal of Western capital. At the same time, the government expended much money and manpower in the erection of "show objects," to impress foreigners, such as convention centers and luxury hotels. Many had to be abandoned, partially completed, for lack of capital and/or manpower. The sacrifices were often not commensurate with the rewards.

In the foreign policy area, as previously indicated, Tito managed to avoid being overthrown by the Soviets and, after the death of Stalin, sought to steer a course that would not obligate him to side with either the Soviet or the Western bloc. At the same time, he managed to exploit his position by getting whatever he could from both sides, and on the whole was fairly successful. He also managed to garner moral recognition by Third World countries, though Yugoslavia did not gain from his personal satisfaction.

The Communists' nationality policy is crucial to understanding the events of 1989–92. Prior to the Tito regime, Yugoslavia was not divided into ethnic units, but in a purported effort to satisfy nationality aspirations and demands, he drew geographic-ethnic lines that left one-third of the Serbs in republics not their own. Yet the regime declared that its solution laid the foundation for brotherhood and unity. Those who purveyed ethnic or religious hostility were to be severely punished. Tito and his comrades must have been worried about possible retribution for the massacres of Serbs in Croatia during World War II. While the dictatorship was not openly defended on those grounds, most Yugoslav citizens understood that a temporary dictatorship of some type was necessary if reprisals were to be avoided.

Contrary to the assertions of some journalists and other observers, ethnic violence among the South Slavs is *not* an ancient phenomenon. There was little such violence prior to the twentieth century. True, some Croatian writers produced anti-Serb diatribes, but these seemed to represent a small minority, and Serbian writers did not reciprocate except to quote Croats. The first concrete signs of hostility between Serbs and Croats came during World War I, when Croats and Slovenes fought in the Austro-Hungarian army against Serbia. Interestingly, Tito was a member of that army, but it is not clear if his hatred of Serbs dated from those years. High-level Serbian Communists have admitted, or at least insisted, that they did not discover his hatred of Serbs until about the mid-1960s.

It is also important to remember that once Tito's Partisans gained control of Croatia near the end of the war, they executed a large number of people, Croats and Serbs, as well as others who were trying to flee from Communist enslavement. Serbs were in no position to retaliate, and never practiced genocide against any people. One of the worst aspects of British meddling in favor of Tito's rise to power occurred when the British military-occupation authorities in Austria forced many of those fleeing from Yugoslavia to return and become subject to the terror of the Partisans.

The Kosovo Albanians, who are not Slavs, are a special case. They are victims for many reasons. They are living on land that is holy ground to Christian Serbs. Historically the Albanians moved there after they had converted to Islam and became vassals of the Ottoman Turks, a vassalage that retarded the realization of their own national identity until late in the nineteenth century. They are victims of the highest birthrate in Europe, which to no small extent explains their poverty. They are the victims of Mussolini's and Hitler's promises that Kosovo could become part of a Great Albania, and of Tito's receptiveness, if not outright encouragement, of large-scale migrations from Albania to Kosovo.

They are the victims of Tito's 1974 constitution, which encouraged them to believe that they could do anything they wanted in Kosovo, which may have contributed to their gross violations of the rights of the Serbian minority. They are the victims, along with the Serbs, of benign neglect by the interwar government of Yugoslavia, and the squandering of large sums of economic development monies from the Tito government, much of which was spent on showcase objects and on the purchase of Serbian properties. They are the victims of hate-Serb propaganda purveyed by professors and textbooks imported from Albania.

It is ironic that the Albanians in Yugoslavia, who in earlier years talked much about minority rights, violated the rights of the Serbian minority in Kosovo on a grand scale, which turned out to be the beginning of the unraveling (giving excuses to Slovenes and Croats along the way) of the failed Communist "solution" to the nationality question. In some circles, Serbia's Milošević was looked upon as the main culprit. It seems safe to conclude, however, that if he had not come to the defense of the Kosovo Serbs, the failure of the Communist policy would not have been long in surfacing, if not in Kosovo then in Croatia and/or Bosnia-Herzegovina.

In any case, once the downfall of the Communist system was no longer in doubt, it was tragic for the peoples of Yugoslavia that recycled Communists acted in a manner that suggested the lack of elemental rationality. Perhaps that should not surprise us, because the Yugoslav Communists were the ones who first propelled the country down that disastrous path of political folly. And the opponents of the Communist system, instead of offering the country an alternative, resorted to narrow ethnic persuasions that led to the end of Yugoslavia.

Moreover, irony of ironies, Yugoslavia's demise was hastened by "friends" (European Community) who offered their good offices, ostensibly to promote a peaceful solution, but, pressured by Germany and Austria to recognize Slovenia and Croatia without delay, they did not wait for the outcome of the negotiations. It is almost unbelievable that several sovereign states in Western Europe, ignoring basic principles of international law and the Charter of the United Nations, took it upon themselves to speed up the destruction of another sovereign European state.

Not to be out of step too long, the United States in April 1992 joined the EC countries in recognizing Slovenia and Croatia, and moved ahead of the EC in extending diplomatic recognition to Bosnia-Herzegovina.

To the Serbs of Montenegro and Serbia this was the final irony. After all of their human and material sacrifices in the Balkan wars as well as in two world wars as allies of the West, they were being told by the European Community and the United States that they should be satisfied to leave nearly three million *once-liberated* compatriots to the whims of other masters.

Lacking other realistic alternatives, Serbia and Montenegro in April 1992 combined to form the Third Yugoslavia. Once a nation nearly twenty-four million strong, Yugoslavia was reduced to less than half that size.

Consequences and uncertainties of untold proportions remain. What paths will the three million Serbs who remain outside the new Yugoslavia follow, particularly those in Croatia and in Bosnia-Herzegovina? What will become of the remnants of the Yugoslav National Army? What is to happen to Macedonia? What of the problem of the Albanian minority in the new Yugoslavia, some two million strong, most of whom live in the historic Serbian "holy land" of Kosovo?

As of this writing (May 1992), there are few answers, and the possibilities for chaos remain enormous.

Epilogue

The haste with which the European Community, subsequently joined by the United States, recognized the secessionist Yugoslav republics—first Slovenia and Croatia, later Bosnia-Hercegovina (can Macedonia be far behind?)—was purportedly to prevent, or at least minimize, bloodshed. Although the motive was laudable, the aim was not attained. As in other civil wars, atrocities, destruction of cultural monuments, and other untoward cruelties were committed by both sides. And truth became the first casualty.

The EC's actions in 1991, at the height of the Yugoslav crisis, left many observers bewildered. The offer of good offices in the hope of promoting a peaceful settlement was understandable. What was difficult to fathom was the rush of EC members, notably Germany and Austria, to announce even *before* any significant negotiations that they intended to recognize the independence of Slovenia and Croatia. Hence would-be peacemakers took sides in advance of any results their supposed good offices were designed to produce.

It should be noted that at the time there was a central government of Yugoslavia in existence to which the EC member states had accredited diplomatic representatives.

Thus, as indicated on a previous page, it is difficult to escape the conclusion that a collection of independent European states actively contributed to the destruction of another sovereign European state—Yugoslavia—in contravention of clearly understood principles of international law and the Charter of the United Nations.

German and Austrian aggressiveness in the drive to recognize Croatia and Slovenia was a painful reminder, especially to the Serbs, of German occupation during two world wars. Could this be revenge, Serbs asked, for the failed Austrian and German efforts in 1914 and 1941 to subdue them? Were the Germans seeking to achieve politically and economically what they had failed to accomplish by military means? Were they once again seeking cheap and secure sources of raw materials and new markets for their massive industrial output? These questions simply will not go away.

Moreover, the EC has been inconsistent. While recognizing the rights of Slovenes and Croats to an independent existence, its members have completely ignored the expressed wishes of the Serbs in Croatia, Bosnia-Herzegovina, and Macedonia.

It must be noted that some EC members, particularly Britain and France, did have reservations. Official sources in Paris suggested that Germany's drive to recognize Slovenia and Croatia was motivated by a desire to create a "Teutonic Belt" from the Baltic to the Adriatic.

The United States wanted, first of all, a peaceful resolution of existing problems between republics, and initially stressed the hope for the continued existence of an integral Yugoslav state. At the same time, it indicated readiness to accept freely expressed decisions of the peoples of Yugoslavia. The actions of the Slovene and Croat leaders in acting unilaterally delayed U.S. recognition.

For better or worse, and it would seem the latter, the American approach was abandoned in favor of the EC preferences.

The secession of Slovenia and Croatia posed many questions. What would happen to Yugoslavia's international obligations, particularly its large foreign debt? What of its many domestic accounts and properties? What of past treaties? Who would be bound by the Versailles treaties, Trianon and St. Germain, which treated the Kingdom of the Serbs, Croats, and Slovenes as the successor state of Serbia? All international agreements to which Serbia was a party had been transferred to Yugoslavia. Technically, it seemed to some international-law experts, Serbia would have a vested interest in *all* the territories Trianon and St. Germain had conveyed to the successor kingdom, which would include the areas of the secessionist republics.

What was to become of the other republics? Before long, plebiscites in Bosnia-Herzegovina and Macedonia indicated a desire for independence, but neither had ever existed as an independent state. The only independent states making up the new Yugoslav state in 1918 were Serbia and Montenegro. In a plebiscite in 1992, Montenegro voted strongly for remaining with Serbia as a part of Yugoslavia.

Only history will tell if the breakaway republics will become viable independent states. Slovenia, with no significant minorities, may have the best chance. Croatia, unless it proves willing to deal realistically with the large Serbian minority which does not want to remain part of that state, may suffer instability for an indefinite time, to say nothing of becoming subservient to Germany and perhaps other European states.

Bosnia-Herzegovina's problems are complex and difficult. Of its population of 4.3 million, the Serbs make up 1.5 million; the Croats, about three-quarters of a million; the Muslims, many of whom think of themselves as Serbs, others as Croats, about 2 million. Prior to the 1971 census there was no such thing as a Muslim ethnic category. In 1991, the Muslims and Croats voted in favor of independence; the Serbs boycotted the vote, indicating that they wanted to remain with Serbia as a part of Yugoslavia. Following EC and U.S. recognition of Bosnia-Herzegovina, the Serbs left the government in Sarajevo and proclaimed a Serbian Bosnian state. This led to bloodshed and destruction on a vast scale, as could have been easily predicted and for which the EC and the U.S., by their hasty recognition, are in no small measure responsible.

Macedonia's situation is in some ways even more complex and fraught with dangers. A part of the Serbian nation in the Middle Ages, Macedonia was known as South Serbia. After liberation from the Turks in the Balkan wars, the largest part became part of the Kingdom of Serbia; smaller parts went to Bulgaria and to Greece. After World War II, Tito made Yugoslav Macedonia a separate republic, and those Serbs who earlier were forced to flee were not allowed to return. Serbian schools were closed; Serbian names were changed, e.g. Jovanovski instead of Jovanović, making it extremely difficult to estimate the number of Serbs there. Albanians make up more than thirty percent. There are some Turks and other minorities. Macedonia's neighbor Bulgaria views Macedonians as Bulgarians. Hence, it recognized Macedonia as a state mainly to keep it from remaining part of Yugoslavia, but made it clear that Macedonians were not being recognized as a separate nationality.

Macedonia's problems were further complicated by the fact that Greece, an EC member, objected strongly to recognition of Macedonia under that name, fearing future claims against areas of northern Greece. Greeks pointed out that Macedonia is part of the Hellenic heritage, and suggested that if there was to be a Macedonian state, it should be known as the Republic of Skopje, from the name of its capital.

To the Serbs of Montenegro and Serbia, after their sacrifices in the Balkan wars and then in two world wars as allies of the West, it seemed a final irony that they were being told by the European Community and by the United States that they should be satisfied to leave one-third of their once-liberated compatriots to the whims of other masters.

The United Nations Security Council sought to rescue the EC and recalcitrant peoples of the former Yugoslavia by offering to provide peacekeeping forces until a political settlement was achieved. Unfortunately, all Western efforts to find a solution to these Yugoslav problems were based on two related but erroneous assumptions: (1) that the boundaries which Tito had imposed were sacrosanct, and (2) that while respecting the wishes and interests of the secessionist republics, the interests and wishes of Serbia and Montenegro could be ignored. If the policy makers had known the history of the Yugoslavs, they would have known that proceeding on the basis of these faulty assumptions was a sure formula for disaster.

Appendix: Population of Yugoslavia in 1991

Bosnia-Herzegovina	4,365,639
Muslims	1,900,000
Serbs	1,450,000
Croats	750,000
Yugoslavs	250,000
Croatia	4,703,941
Croats	3,500,000
Serbs	700,000
Yugoslavs	400,000
Macedonia	2,033,964
Macedonians	1,300,000
Albanians	425,000
Serbs	100,000(?)
Turks	100,000
Bulgars	50,000(?)
Gypsies	40,000

Montenegro	616,327
Serbs	550,000
Yugoslavs	40,000
Albanians	25,000

Serbia	9,721,177
Serbia Proper	5,753,825
Serbs	5,500,000
Muslims	125,000
Gypsies	50,000
Croats	40,000
Kosovo	1,954,747
Albanians	1,630,000
Serbs	250,000
Muslims	40,000
Gypsies	30,000
Vojvodina	2,012,605
Serbs	1,400,000
Hungarians	450,000
Croats	100,000
Romanians	50,000

Slovenia	1,974,839
Slovenes	1,800,000
Croats	60,000
Serbs	50,000
Yugoslavs	50,000

From *Yugoslav Survey, XXXII* (March, 1990–91), 5. The numbers for republics and autonomous provinces are actual; the numbers for the population groups are estimates.

Selected Bibliography

Adelman, Jonathan, ed. *Superpowers and Revolution*. New York: Praeger, 1986.

Alexander, Stella. *Church and State in Yugoslavia Since 1945*. Cambridge: Cambridge University Press, 1979.

Armstrong, Hamilton Fish. *Tito and Goliath*. New York: Macmillan, 1951.

Balfour, Neil, and Sally Mackay. *Paul of Yugoslavia: Britain's Maligned Friend*. London: Hamish Hamilton, 1980.

Banac, Ivo. *The National Question in Yugoslavia: Origins, History, Politics*. Ithaca, NY: Cornell University Press, 1984.

Beloff, Nora. *Tito's Flawed Legacy: Yugoslavia & the West Since 1939*. Boulder, CO: Westview Press, 1985.

Bičanić, Rudolf. *Economic Policy in Socialist Yugoslavia*. Cambridge: Cambridge University Press, 1973.

Ciano, Galeazzo. *The Ciano Diaries, 1939–1943*. Edited by Hugh Gibson. New York: Doubleday, 1946.

Deakin, F. W. D. *The Embattled Mountain*. London: Oxford University Press, 1971.

Djilas, Aleksa. *The Contested Country: Yugoslav Unity and Communist Revolution, 1919–1953*. Cambridge, MA: Harvard University Press, 1991.

Djilas, Milovan. *Memoir of a Revolutionary*. New York: Harcourt Brace Jovanovich, 1973.

———. *The New Class: An Analysis of the Communist System*. New York: Praeger, 1957.

———. *Rise and Fall*. New York: Harcourt Brace Jovanovich, 1985.

———. *Tito: The Story from the Inside*. New York: Harcourt Brace Jovanovich, 1981.

———. *Wartime*. New York: Harcourt Brace Jovanovich, 1977.

Djordjević, Dimitrije, ed. *The Creation of Yugoslavia, 1914–1918*. Santa Barbara, CA: Clio Books, 1980.

Dragnich, Alex N. *The Development of Parliamentary Government in Serbia*. Boulder, CO: East European Monographs, 1978.

———. *The First Yugoslavia: Search for a Viable Political System*. Stanford: Hoover Institution Press, 1983.

———. *Serbia, Nikola Pašić, and Yugoslavia*. New Brunswick, NJ: Rutgers University Press, 1974.

———. *Tito's Promised Land: Yugoslavia*. New Brunswick, NJ: Rutgers University Press, 1954.

Dragnich, Alex N., and Slavko Todorovich. *The Saga of Kosovo: Focus on Serbian-Albanian Relations*. Boulder, CO: East European Monographs, 1984.

Henderson, Neville. *Water Under the Bridges*. London: Hodder & Stoughton, 1945.

Hoptner, Jacob. *Yugoslavia in Crisis, 1934–1941*. New York: Columbia University Press, 1962.

Ivanović, Vane. *LX: Memoirs of a Yugoslav*. New York: Harcourt Brace Jovanovich, 1977.

Jukić, Ilija. *The Fall of Yugoslavia*. New York: Harcourt Brace Jovanovich, 1974.

Lees, Michael. *The Rape of Serbia: The British Role in Tito's Grab for Power, 1943–1944*. New York: Harcourt Brace Jovanovich, 1990.

Macek, Vladko. *In the Struggle for Freedom*. University Park: Pennsylvania University Press, 1957.

Maclean, Fitzroy. *Eastern Approaches*. London: Cape, 1949.

Martin, David. *The Web of Disinformation: Churchill's Yugoslav Blunder*. New York: Harcourt Brace Jovanovich, 1990.

Pavlowitch, Steven K. *The Improbable Survivor*. London: C. Hurst, 1988.

Petrovich, Michael B. *A History of Modern Serbia*. 2 vols. New York: Harcourt Brace Jovanovich, 1976.

Roberts, Walter R. *Tito, Mihailovich, and the Allies, 1941–1945*. New Brunswick, NJ: Rutgers University Press, 1973.

Rootham, Jasper. *Miss Fire: The Chronicles of a British Mission to Mihailovich, 1943–1944*. London: Chatto & Windus, 1946.

Rothenberg, Gunther E. *The Military Border in Croatia 1740–1881*. Chicago: University of Chicago Press, 1966.

Rubinstein, Alvin Z. *Yugoslavia and the Non-Aligned World*. Princeton, NJ: Princeton University Press, 1970.

Rusinow, Dennison. *The Yugoslav Experiment: 1948–1974*. Berkeley, CA: University of California Press, 1977.

Sirc, Ljubo. *The Yugoslav Economy Under Self-Management*. London: Macmillan, 1979.

Ulam, Adam. *Tito and the Cominform*. Cambridge, MA: Harvard University Press, 1952.

Vucinich, Wayne, ed. *At the Brink of War and Peace*. Stanford, CA: Stanford University Press, 1983.

West, Rebecca. *Black Lamb and Gray Falcon: A Journey through Yugoslavia*. 2 vols. New York: Penguin, 1982.

Index